THE SUN BETRAYED

A Report on the Corporate Seizure of U.S. Solar Energy Development

THE SUN BETRAYED

A Report on the Corporate Seizure of U.S. Solar Energy Development

by
Ray Reece

South End Press

Cover Design by Carlene Brady

Library of Congress Catalog Card No: 79-66992
ISBN 0-89608-071-4
ISBN 0-89608-072-2

Printed by Maple Vail, York, Pa. U.S.A.

Typesetting, design and paste up done by the
South End Press collective.

**South End Press, Box 68, Astor Station
Boston, Ma. 02123**

For My Parents

ACKNOWLEDGEMENTS

Thanks to the following people for support and assistance in the preparation of this book: Howard Bray, the Fund for Investigative Journalism; Jim Hightower, *The Texas Observer*; Michael Albert and the South End Press collective; Ken Bossong and Scott Denman, Citizens Energy Project; Fred Branfman, California Public Policy Center; Kirkpatrick Sale; Daryl Janes; Ronnie Dugger; Gail Vittori; Philip Russell; Carol Rouse; James Benson; Don Elmer; Jim Piper; Harold Hay; Dean Rindy; Mark Parsons; Mark Naison; Ken Smith; Susan and Bill Yanda; James Ridgeway; Robert F. King; Russel Smith; Susan Thayer; Howard Scoggins. Thanks especially to Ave Bonar, whose patience and stamina helped make the book possible.

TABLE OF CONTENTS

"About one-fifth of all energy used around the world now comes from solar resources: wind power, water power, biomass, and direct sunlight. By the year 2000, such renewable energy sources could provide 40 percent of the global energy budget; by 2025, humanity could obtain 75 percent of its energy from solar resources.... Every essential feature of the proposed solar transition has already proven technically viable; if the 50-year timetable is not met, the roadblocks will have been political—not technical."

—Denis Hayes, *Rays of Hope*

"As the problems identified are resolved...solar energy may account for one or two percent of the nation's electric power capacity by the year 2000 and about two percent of total energy use."

—Piet Bos, Electric Power Research Institute

"Our conclusion is that with a strong national commitment to accelerated solar development and use, it should be possible to derive a quarter of U.S. energy from solar by the year 2000. For the year 2020 and beyond, it is now possible to speak hopefully, and unblushingly, of the United States becoming a solar society."

—White House Council on Environmental Quality

"We can't afford the luxuries of basing energy policies on fantasy, such as solar energy...or some perpetual motion machine that will bring us out of our problems. As supplies of natural gas and petroleum dwindle, this nation will become dependent on coal and nuclear fission. Like it or not, the public must be made aware that there is no choice."

—Rep. Mike McCormack, U.S. Congress

AUTHOR'S PREFACE

In Austin, Texas in 1975—as in countless other communities across the United States—there coexisted two distinct but overlapping groups of people whose principal bond was a sense of alienation from mainstream corporate America. On the one side was clustered a loose and motley amalgam of generally well educated, highly talented scientists, technicians, architects, and ecologists who, for a variety of reasons, had chosen not to spend their talents and their lives in salaried pursuit of The Great American Suburban Dream. Instead they had joined that ragged subclass born of the turbulence of the late 1960's, tempered in the struggle against the war in Vietnam, and finally tagged the "counterculture." In place of lobster and martinis for lunch in paneled private clubs, they dined on beansprouts and passed a joint among friends in the park. Instead of conducting experiments in corporate laboratories or designing banks and office spires or washing young brains in college classrooms, they worked as carpenters and co-op managers, health-food chefs and taxi-drivers. They were—relative to their skills, education, and deftness at solving problems—underemployed, underutilized, and undercompensated.

1

On the other side of the corporate mainstream stood a slightly more conventional but no less restive group of young to middle aged aspiring entrepreneurs who still believed, as their parents, teachers, and politicians had taught them to believe, that they might one day operate their own independent small business and technological enterprises. They believed this despite increasingly massive evidence to the contrary, despite the vanquishing of small private research labs and manufacturing concerns by Exxon, Westinghouse, and Texas Instruments, despite the disappearance even of independent auto repair shops and construction firms. Some of these dreamers worked in the plants and drafting rooms of the very corporations which had foreclosed their dreams, but they dreamed nonetheless, saving a little of each month's salary against that day when they would break away on their own.

Both of these groups, in Austin as in other American cities, developed what seemed at the time a rather odd fascination with the "energy crisis" which settled on the country in the winter of 1973. They pricked up their ears through the spring and summer of 1974 as the U.S. Congress debated a series of relatively drastic legislative measures aimed at bringing the nation out of the "crisis." Among those measures was the Energy Research and Development Act, passed in the autumn of 1974 with a budget of millions of dollars to be invested in a nationwide search for unconventional alternatives to petroleum energy resources. Among the alternatives to be explored were solar, wind, and biomass technologies—each of which was presumably best deployed in small-scale decentralized applications. Among the clauses in the 1974 legislation was one instructing the federal energy bureaucracy to award a significant proportion of its new R&D contracts to small business.

During the period of these congressional initiatives, I was working for an Austin-based architecture magazine. The nature of my work had placed me in a position to observe both groups of proto-independent technologists and entrepreneurs, those in the "counterculture" and those on hold in nine-to-five jobs. I watched them both conclude that the federal initiatives of 1974 might open the way for them to apply their latent skills and interests in a manner more gratifying than either group had expected from the federal government. I watched them organize shoe-string companies and research teams that conjured up, mulled, and polished ideas for small-scale renewable energy devices and projects ranging from attached solar greenhouses and collector systems to catalytic methane gas converters and solar-thermal electric machines. Throughout the better part of 1975 and early 1976, these ideas were laboriously translated into funding proposals and mailed to the Energy Research and Development Administration in Washington, D.C. There was hope in the air around Austin that year.

I watched, too, as proposals from both groups started coming back unfunded, some apparently even unread by officials in Washington. Not just skimpy ill-defined ideas were rejected, but proposals for innovations so simple and yet so brilliant in concept as to be virtually failsafe, effective, economical. The rejections mounted through the fall and winter of 1976 with a dreary consistency that soon turned the hope and gusto of the previous year to dismay, disillusion, and renewed contempt for the government. Many of the enterprises formed by the hopeful solar independents had collapsed by the following spring, the hopefuls themselves returning to salaried jobs and taxi stands.

Meanwhile, it had come to my attention that millions upon millions of those federal solar research dollars were being doled out to precisely the corporations and companion institutions viewed by many in the small-scale solar community as having essentially fabricated or at least compounded the nation's "energy crisis" in the first place. This was hard to swallow: by what miracle had the Exxons and Lockheeds of the land become the fount of innovative alternative energy technologies for the federal government? What were the corporations producing in this area that was so superior to the concepts developed by the small R&D groups whom I had observed? If the corporate solar concepts *weren't* in fact superior—more inventive and economical, more appropriate to the nature of the problems involved—then why were the corporations being funded to the near-total exclusion of the smaller firms?

In pursuit of the answers to these questions, I enlisted the support of the biweekly *Texas Observer* in securing a grant from The Fund for Investigative Journalism and, in June, 1977, embarked upon the task of researching and writing a series of what I assumed would be two or three articles for the *Observer*. By August of that year, it was apparent that a book would be required to tell the story of what I was finding in my research: a story of good intentions driven astray by the politics of corporate hegemony—a story of deceit, vested interest, and collusion in the highest echelons of U.S. industry and government. That story follows.

Before the telling of it, however, an observation concerning the two groups of small-scale technologists cum entrepreneurs who are cited above. Many of these risk-takers in the solar field, particularly those outside the "counterculture," are, in a word, capitalists. Their attraction to solar energy has little or nothing to do with the socioeconomic, environmental, and moral implications of a widespread deployment of small-scale renewable energy technologies. Their attraction to these new technologies has rather to do with what they perceive as a rare opportunity to combine their taste for independence with an otherwise traditional desire for success in the marketplace: they want to be rich and they want to be noticed. Some of them, as events have shown, are ultimately willing to sell their solar enterprises to the highest corporate bidders, while others hope to build their enterprises to major corporate dimensions themselves. In this respect, driven above all by pecuniary self-interest and ego, they are different only in degree, not in kind, from the U.S. corporate oligopolists who have done so much to help snuff out independent small business in America.

Yet that degree of difference is a critical one, for these entrepreneurs do indeed possess a number of the characteristics historically associated with their breed. They are highly inventive. They are locally rooted, with a subsequent sense of community responsibility, and they are willing both to sacrifice and to compete for their due position of prominence in the marketplace. These happen to be measurably valuable attributes at the present stage of solar energy development in the U.S., and so it is that the book which follows is a relatively uncritical, undifferentiating champion of the small entrepreneurs in their struggle for market position—against poor odds made poorer by the undemocratic behavior of their government. It is only in the final chapter, where questions are raised concerning the necessity for radical adjustments in the nation's socioeconomic and political order, that anything less than sympathy and respect for the small solar energy entrepreneur is advanced.

A closing word of caution to the reader: in presenting the following history of corporate-government collusion in the making of a solar energy market, I have found it necessary to include a series of mini-catalogs of institutions and persons who have played the starring roles in that collusion. Most of these interwoven "catalogs" appear in Chapters II and III, and less indulgent readers may want to skip or skim the "catalogs" en route to zippier material. Other readers may want to *study* the "catalogs" by way of grasping the almost unbelievable pattern of interlocks revealed therein. Finally, attentive readers will notice that the text of the book is free of footnotes identifying sources and references. This material, arranged by chapter in order of citation, has been placed at the end of the book.

Find out who is Energy Secretary!

(Bulletin: As this book goes to press in early August, 1979, there is news of two very recent developments on the national energy front that bear implicitly upon some of the material in the book. One development is the forthcoming replacement of Energy Secretary James B. Schlesinger by Charles W. Duncan, Jr., currently serving as Deputy Secretary of Defense in the Carter administration. The second development is the appointment of Denis Hayes, chairman of the grass-roots "Solar Lobby," as director of the Solar Energy Research Institute in Golden, Colorado, a national solar R&D facility funded by the Department of Energy.

While dramatic on the surface, neither of these two developments is likely to have a significant near-term impact on the dubious state of U.S. solar energy affairs as depicted in the following pages. This is especially true of President Carter's appointment of Charles Duncan as Secretary of Energy. If anything, the Duncan appointment simply assures that the corporate policies and programs addressed in this book will continue more systematically and perhaps more efficiently than under Schlesinger.

Indeed, like Schlesinger himself, Duncan is not only a proponent of accelerated nuclear energy and synthetic fuels development, he comes to Energy from the Department of Defense, where he has made a name for himself as a "tough manager, a businessman through and through," according to Pentagon observers quoted by Knight News Service. This underscores the ill-disguised and ominous relationship between U.S. "defense" and "energy" objectives discussed in chapter I below.

In addition to his "management" experience at the Pentagon, Duncan brings to his post at the Department of Energy a set of impeccable corporate credentials. He is, for one thing, a scion of the Houston family who built the Duncan food empire. He is also a former president of Coca-Cola, and chairman of Rotan Mosle, an international investment company with significant holdings in oil, gas, coal, and uranium corporations—not to mention investments in the corporate solar market. With regard to his preferences for this or that energy resource: "He is from Houston," said a Pentagon spokesman quoted by *The Energy Daily* (July 23, 1979), "and I believe that's the headquarters of the oil industry." Schlesinger himself remarked, at a Washington news conference shortly after Duncan's appointment, that "it will be precisely the same game with different players."

If not for the above realities, the appointment of Denis Hayes to direct the Solar Energy Research Institute would be a piece of unalloyed good news, at least in view of Hayes' reputation as a champion of grass-roots participation in the development of U.S. solar energy policy. The problem is that Duncan will be Hayes' boss.)

I

U.S. Energy Policy Since 1971:
Toward the Survival of Global Capitalism

Dusk is falling on a January day in Crystal City, Texas, and the wind from the west is cold. It blows unfettered across the dry sage and the sandy gulches of a plain growing dark beneath the mauve and the black-laced orange of the sky. It howls through the town and along a ravine to a clump of barren mesquite trees two miles east, whistles in the limbs and bites at the fingers of a dark-skinned boy who gathers an armload of fresh-cut mesquite and stacks it on a pile in the rear of a battered pick-up truck. Beside him his father is wrapping an axe and a pair of handsaws in burlap cloth. His fingers too have gone numb in the wind, fingers blunted by a lifetime of work picking onions and spinach in other men's fields. He drops the tools with a clunk in the bed of the truck and motions for the boy to get inside. Their doors clang shut. The man starts the engine and drives without speaking toward the lights of the town, clutching the wheel against the heaving of his truck in the wind. The boy next to him cups his hands to his mouth, trying to warm them with his breath. Approaching the town, the man turns onto a narrow street and stops the truck in the packed earth yard of a flimsy wooden shack.

He and the boy climb out of the truck, facing the wind-shrill night again, and scoop up loads of twisted mesquite and enter the shack where a silver-haired woman is waiting in a chair with a Navajo blanket across her legs. She is glad to see them. She rises, throwing the blanket around her shoulders, and helps them stack the wood adjacent to a black Army camp-stove whose shiny blue exhaust pipe ascends through a hole in the roof. As she works, the woman's breath forms little clouds of vapor in the air. The man grunts in with another heap of logs, stacking some and using the others to ignite a fire in the squat iron stove. It flares and crackles to life. He shuts the grate and stands, kissing the woman's furrowed cheek, and turns to leave to make other stops, other deliveries to family and friends. The boy trudges after him through the door, and the woman smiles as their truck pulls away, rubbing her hands atop the black stove. She glances at another stove looming large and porcelain white in a corner of the room. Not long ago it had burned the gas that cooked her food and boiled her water and warmed her house, easing the ache of winter in her joints. Now it is dead. It is cold and useless save as a table on which to store utensils and jars and boxes of food. Her smile has vanished as she walks to the stove and removes a cloth from a bucket of water and stares at the film of ice on the surface. She cracks the ice with a small copper pan. She dips up some water and returns to the black stove, setting the water on the stove to heat so that later she can brew a cup of tea. The west wind sings through the cracks in her walls, and she draws the blanket as tightly about her as she can.

This is true. The woman's name is Tirsa Gonzalez, and she is one of 8000 residents of Crystal City whose supply of natural gas was shut off in September, 1977, by Lo-Vaca Gathering Company, a pipeline monopoly in south central Texas. For the five years prior to that time, in violation of a 20-year contract signed by Lo-Vaca in 1963, the company had been raising its prices above the contract rate of 36 cents per thousand cubic feet (mcf) of gas. By early 1975, the rate was $1.87 per mcf, and the Crystal City government had lost a number of expensive court battles, including an appeal to the state gas regulatory commission, to force Lo-Vaca to honor its original commitment. So, while awaiting the outcome of further lawsuits, angry municipal officials simply stopped paying Lo-Vaca's higher charges, reverting instead to the price in the contract of 1963.

Two years later, with the town more than $800,000 "in arrears" to Lo-Vaca, the gas stopped flowing to the people of Crystal City, and the rest of the people of the United States were afforded a glimpse of their own not-distant future. They were also afforded a rare opportunity to begin preparing for that future with solar energy. Indeed, they had a chance through an agency of their government to transform the humiliation of Crystal City—the town whose gas was cut off—into a showcase of new technologies and energy systems vastly more appropriate to a future which assuredly will not include an abundance of natural gas or of any other depletable fuel, particularly oil and uranium. Crystal City was tailor-made for such an experiment in survival. The people there are poor, subsisting on a median family income of $3400 per annum, most of it earned as laborers on profitable corporate farms in the area. Many of the workers are unemployed for six and nine months of the year. The local government is progressive, devoted to its people, and committed to building an indigenous economic base such as might be provided by a "cottage industry" turning out small-scale renewable energy systems for local application. There awaits in nearby Austin, furthermore, a community of young "appropriate technologists" eager to help design and implement the foundation for a solar-based "cottage industry" in places like Crystal City.

Two of those young technologists are Pliny and Daria Fisk, former professors of architecture at the University of Texas at Austin and co-directors of the Center for Maximum Potential Building Systems, a non-profit organization with associates and supporters around the world. In late September, 1977, the Fisks submitted to the federal government an emergency grant request for $80,000 to help launch a crash-program of solar installations and technical training for Crystal City residents now confronting a winter without gas. Backed by the mayor and other officials of Crystal City, the Fisks had hoped that the Energy

Research and Development Administration (ERDA), to whom they had submitted their proposal, might act quickly enough for them to have their energy program well underway by the onset of hard weather in late December, 1977. (The Fisks were not relying on solar installations alone. As an interim measure, they had planned to equip a larger number of homes with small efficient oil-fired stoves to be assembled in a makeshift local "factory." They had further planned to involve the residents of Crystal City in a hurried campaign to insulate their own houses, most of which were as ill-protected from the weather as Tirsa Gonzalez' little dwelling.) What the Fisks needed most from the government was haste, good faith, a dash of imagination, and a willingness to waive red tape in favor of assistance to indigent Americans threatened by disaster and wishing ultimately to assist themselves.

Instead the Fisks and the people both of Crystal City and the rest of the nation were subjected to a laughless comedy of errors, internecine bickering, and bureaucratic myopia. Shortly after the news broke that Lo-Vaca had terminated gas service to a city of 8000 people, U.S. Senator Edward Kennedy released a statement that the Community Services Administration had promised to grant Crystal City $160,000 to help defray its "debt" to the pipeline company. This brought squeals of protest from officials of other U.S. townships, particularly those in Lo-Vaca's south Texas service area, who had slavishly remitted the ever-higher fuel payments imposed on them by utility executives quite content to shear as many sheep as would stand for the shearing. Due to this wave of protest, the CSA was forced to withdraw its offer at about the time that ERDA rejected the Fisks' proposal for a crash-program of solar installations and community self-help in Crystal City. As a token of its concern, however, and perhaps in view of Mexican-American voting strength, the federal government did make available through CSA a total of $310,000 in assistance to Crystal City residents.

Of this amount the Fisks were granted $3000 by federal officials in Crystal City to conduct a series of small-stove workshops and to build a demonstration solar collector. The Fisks, in keeping with their reputation, stretched those funds to the limit, setting up in a local warehouse a combination mini-factory and training center for residents wishing to build and maintain their own "alternative energy systems." By the time their money was exhausted in February, 1978, the Fisks had taught more than 200 farm laborers how to construct a variety of autonomous home-scale energy devices, including solar ovens and stoves designed to burn recycled crankcase oil. Some of their "students," moreover, had built and installed a solar collector on a local house which became an instant celebrity, attracting people from all over town to come wash their hands in water heated entirely by the sun on a sub-freezing February day. Without additional funds, however, the Fisks were unable to parlay their successful demonstration into the city-wide "energy revolution" which could easily have followed at that point. Further appeals to granting agencies had already been rebuffed, so the Fisks headed back to Austin with the bittersweet taste in their mouths of a job gamely started but so far from finished that it verges on failure. "I have to admit I'm disappointed," said Pliny Fisk in an interview later. "The collector was working. The people saw it working and they clapped their hands as if they'd just discovered fire." He threw up his arms in frustration.

Meanwhile, the remaining $307,000 in federal assistance to Crystal City was being spent according to government mandate on such "emergency items" as butane gas, causing the price of butane from local suppliers to double in 30 days, which made it too expensive for poorer residents. Also purchased with government funds were electric blankets and hot plates, as well as Army campstoves of the sort now employed by Tirsa Gonzalez, fueled with mesquite logs scavenged from the paltry countryside around Crystal City. Tirsa Gonzalez might have had solar energy. She doesn't, and Crystal City doesn't, and she is suffering, and she doesn't know why. "I paid all my bills," she told an interviewer in January, 1978. "I've got the receipts to prove it. So tell me, why did they shut off the gas? What did we do to deserve such a thing?" Tears began to well in her soft dark eyes. "It's so sad," she murmured, "to be old and cold."

* * * * *

What didn't happen in Crystal City in the fall of 1977 is a tiny microcosm of what hasn't happened to solar energy in the United States. It is now disturbingly apparent, in fact, that the makers of U.S. energy policy have not the slightest intention of a rapid and effective deployment of solar power and its kindred beneficent technologies. On the contrary, an overhauled national schedule calls for a massive acceleration of coal and nuclear power as replacements for declining stocks of oil and natural gas. These will be stretched by exploitation of off-shore reserves and oil blasted from shale, as well as by synthetic fuels from coal. Such an arrangement will allegedly permit industrial America to expand its energy appetite by 3 percent a year (down from 4.5 percent in 1977), with "conservation" expected to account for 16 percent of the annual increment by 1985.

If solar energy is to have a function at all, it will be what top-level strategists call a "supplemental" one through at least the remainder of this century, and it will be controlled by large corporations and utilities. The solar function will be shaped, moreover, so as not to inhibit a projected growth in electrical power generation from 14 percent of the nation's total energy supply—the current figure—to as much as 60 percent by the year 2000. The additional megawatts will come primarily from 450 to 600 new coal-fired power plants and an equal number of nuclear plants, scores of which, because of predicted shortfalls in virgin uranium supplies, will inevitably feature the deadly plutonium breeder reactor. (Jimmy Carter's much-publicized opposition to the Clinch River breeder project is merely a palliative to environmentalists and perhaps to his own conscience. It is not a threat to the breeder program.)

This, in crude outline, is the "National Energy Plan" announced by Carter on April 20, 1977, and mauled by Congress enroute to passage in December, 1978. The plan has been skewered by a legion of critics, among the most astute of whom is socialist-ecologist Barry Commoner. He points out that the nuclear-electric heart of the plan, due to require a capital investment of $600 billion or more, will lock the nation into a nuclear future while impoverishing the taxpayer, starving social welfare programs, heightening inflation and unemployment, threatening human survival, and relegating solar energy to a long-term role on the periphery of the U.S. energy grid. Besides which, said Commoner in a recent address, the Carter plan "represents the biggest intensification of control over the economy of the U.S. by big corporations that we have seen in our lifetime."

Commoner hit the bull's-eye. A study of the matter reveals that America's megacorporations, led by utilities and oil companies with academic allies, have dominated every important stage in the revision of U.S. energy policy to blunt the potential of solar power. Indeed, the study reveals something akin to an international technocratic conspiracy of giant corporations and giant government headquartered chiefly in Washington D.C. and Wall Street, with field operations strung from Palo Alto and Chicago to Houston, London, and Brussels.

Since 1971, the principals in this mission have not only evolved a coherent international energy policy—in concert with their views of American priorities in foreign policy, defense policy, and fiscal/monetary policy—they have effected that policy through a variety of means not excluding familiar dirty tricks. They have loaded federal agencies and advisory committees with corporate loyalists, conducted illegal secret meetings and political campaigns, falsified energy data, and punished defectors by firing them from government jobs. In the process, they have spun a web of interlocks between corporate and government entities which defies breaching by those Americans, including advocates of a solar alternative, who would have the nation take an energy path less dependent on the senile promise of coal and nuclear power. (A paradigm of this system of interlocks is the new Department of Energy, headed by James Schlesinger, whose preparation for the job included stints as director of the CIA, director of the Atomic Energy Commission, and Secretary of Defense under Nixon and Ford. Dr. Schlesinger is also a former employee of International Telephone and Telegraph Company, itself a prominent force in the development of America's new energy strategy, as well as recipient of millions of dollars in federal research and development money. But that, compared to Schlesinger's mandarin role on behalf of the entire corporate class, is beside the point.)

Observers have suspected for several years that the federal government, despite much rhetoric to the contrary, is not committed to a timely activation of solar and wind energy sources. One early omen was a trio of half-million-dollar studies funded by the National Science Foundation (NSF) in 1974. The studies were conducted by Westinghouse, General Electric, and TRW, Inc. (a major electronics firm), and what they concluded was that solar energy in the year 2000 might contribute between 1.6 percent and 3.5 percent of the nation's total energy supply. This confounded solar technologists whose own arithmetic showed a possible solar input of at least 25 percent and as much as 40 percent by 2000. (A subsequent study in 1978 by the White House Council on Environmental Quality confirmed the 25 percent figure, assuming an all-out push for solar.)

At any rate, an expanded program of solar and wind research was supposed to issue from the Solar Division of the Energy Research and Development Administration (ERDA), established by Congress in January, 1975, and a serious chunk of that program, perhaps up to 50 percent, was expected to be shared by thousands of American small businesses and independent researchers who seemed particularly well suited to the development of a small-scale technology like solar power. These expectations were quickly smothered in a flood of ERDA solar funds to large aerospace and energy corporations, utilities, and favored universities with a predilection for high-technology research. In 1976, for example, out of a total of $94 million spent directly on solar applications, small business received only $6.7 million in prime contracts, or 7.1 percent of the total. (Even that figure is severely clouded by the government's definition of a "small business" as one employing up to 1000 people.) The rest of the solar money went to other government agencies, big universities, and corporations like General Electric ($2.8 million) and Martin Marietta ($3.5 million).

This trend was so pronounced from the very first days of ERDA's operation that hearings were conducted by the Senate Select Committee on Small Business in 1975 to determine whether small-scale solar energy was being suppressed or misdirected by the very bureau assigned to promote it. The evidence at those hearings suggested strongly that such a suppression was indeed underway, and later evidence erased any lingering doubts. ERDA's vision of solar energy was a Buck Rogers vision of fancy technology to be developed by the same corporations, at the same rate of profit, who had earlier brought us nuclear bombs and power plants, saturn rockets, and leaky off-shore oil rigs. (The absorption of ERDA into Schlesinger's Department of Energy has produced nothing more than a refinement of the same inelegant vision. Not only is Schlesinger an ardent devotee of nuclear power and related components of America's military industrial complex—as documented in the pages to follow—his new Department of Energy has made a place for virtually every ranking employee of ERDA, itself the heir to long-entrenched programs and officials with whom Schlesinger worked at the Atomic Energy Commission. The new department has also inherited most of ERDA's civilian advisory committees, including one with direct and active ties to the very pinnacle of the corporate elite. Much of the following discussion, therefore, and especially those parts with reference to issues or programs which originated prior to October, 1977, will use the term "ERDA" more or less interchangeably with "Department of Energy." Recent developments at the Department of Energy, including a 1978 "Domestic Policy Review" of the federal solar program, are summarized in chapter XII.

It is worth noting that the issue involved here is not one simply of corporate porkbarrels nor of conflicts of interest nor of abject discrimination against small business. The issue turns on nothing less than a confrontation between two opposing and mutually exclusive perceptions of the future of western industrial society. One of those perceptions is the corporate ERDA vision sketched above. It implies an energy future based heavily on coal and nuclear technologies, "supplemented" by solar applications which are wastefully complex and expensive. Among the certain effects of such a vision, at best, are environmental damage, political autocracy, and economic pressures which may well split the fabric of our social order.

The second perception has been advanced by a growing number of solar scientists, physicists, and economists who argue that global well-being, if not the survival of humanity, dictates an immediate turning from the path of conventional industrial technology—characterized by mammoth refineries and power plants serving millions of customers each—toward a prompt and resolute emplacement of "appropriate technology" scaled for use by small communities and individual units of production (as the Fisks had planned in Crystal City). Perhaps the most authoritative current spokeman for the latter vision is Amory Lovins, a young physicist who jolted the international energy club with an essay in *Foreign Affairs* (October, 1976) entitled "Energy Strategy: The Road Not Taken?"

The great force of Lovins' argument derives from his skillful use of the very data and econometric models of the corporate planners themselves to demonstrate that their "solution" to the energy crisis—which Lovins calls the "hard path"—is not only unnecessary but unworkable and fraught with apocalyptic consequences. The capital costs alone of installing the energy facilities required by the "hard path" would total more than $1 trillion by 1985. This amounts to three fourths of all domestic capital expected to be available by then, compared to one fourth of annual domestic capital now absorbed by energy investments. Such an outlay, as Commoner suggests, would strip the economy of funds for virtually any other purpose, and, quoting Lovins: "A powerful political response can be expected."

Lovins opposes the corporate strategy with what he calls the "soft path." This begins with a determined program of energy conservation which could halve our national energy demand by the year 2000 without degrading our "standard of living." It would exact a tiny fraction of the projected cost of developing additional fossil resources, thus providing the capital to replace the fossil-based system itself with a new decentralized energy system based on solar and wind technologies. The social returns inherent in such a transition would be manifold, increasing in an exponential ratio with the distance and pace at which "hard" technologies and economic arrangements are left behind. World peace would be enhanced, for one thing, due to a curbing of the gluttonous appetite of industrial states for diminishing natural resources. Corollary payoffs would include a revitalized emphasis on community life and institutions, community control over resources, permanent protection of the environment, and a long-term proliferation of "employment" opportunities based on the labor-intensive character of "soft-path" farming methods and modes of commodity production.

There is little question that the "soft path" is technically and economically feasible. Aside from Lovins' own exhaustive calculations (which finally persuaded a distinguished nuclear physicist named Hans Bethe to accept the technical substance of Lovins' argument), numerous other authorities have produced insuperable documentation. One of these is Robert H. Murray, a physicist and mathematician who helped prepare an encyclopedic study of global solar prospects for the International Center for Integrative Systems in New York. "Solar energy technology," writes Murray, "can make direct thermal energy available at any temperature between ambient and 3800 degrees centigrade. It can produce temperatures above anything required by industry. Therefore, all our present energy functions could be met with solar resources, although complete reliance on solar energy would probably not be realized before the year 2025 to 2050."

Here lies the importance of the confrontation between the "hard" and "soft" energy paths. The "soft path," were it undertaken tomorrow, would require some 50 or 60 years to implement, during which time we would have to depend on remaining supplies of conventional fuels. If we consume those fuels in vain pursuit of the "hard path," they will not be available when it is time to rescue ourselves with an alternative system. "Most energy policy," writes Denis Hayes in *Rays of Hope,* "continues to be framed as though it were addressing a problem that our grandchildren will inherit. But the energy crisis is *our* crisis. Oil and natural gas are our principal means of bridging today with tomorrow, and we are burning our bridges."

It would thus seem irrational for America's corporate energy strategists to blunder farther down the "hard path" which they have clearly chosen. From their point of view, however, the "hard path" *is* the rational path, for they are constrained by the needs of the corporate class itself (as distinct from society in general) to subscribe to a different system of "rationality" than the ordinary person—and the corporate class is afflicted with a swarm of troubles. Indeed, following on the close of the Vietnam crime in 1972, the chieftains of the corporate elite confronted a tangle of crises and dislocations which loomed as threats to their entire system: (1) the loss of the war itself, slamming the door on 25 years of freewheeling Cold War expansion of corporate foreign markets; (2) stiffening resistance by the Arab countries to monopoly domination of their oil fields, coupled with increasing U.S. dependence on oil imports; (3) substantial deficits in the international balance of payments; (4) mounting inflation and unemployment in the United States; (5) a weakening of the dollar in global markets so acute that the dollar was devalued in 1973; (6) sweeping cutbacks in federal contracts to aerospace and defense corporations; (7) a 10-year lag in the U.S. nuclear electrification program, the blame for which was (wrongly) ascribed to a mass environmental movement that struck the corporations as being out of control.

These and other factors have driven the corporate elite since 1971, as mentioned above, to introduce an aggressive new masterplan tying energy policy not to human survival but to U.S. economic growth, "national security," and foreign trade considerations. Not that such items haven't always been linked in the modern corporate mind: at least since World War II, when combat production salvaged the capitalist order from a Great Depression which had kept the system on its knees for 12 years, the corporate leadership has perceived and taken advantage of a useful symbiosis between U.S. "defense" needs and the health of the corporate economy. That symbiosis was broadened through the post-war Marshall Plan, enabling corporate America to buy up much of Western Europe and setting the stage for a truly international corporate dominion. The present malfunctions of the U.S. economy, however, which may be more serious than those at work in the Wall Street bust of 1929, have infused the symbiosis with a fresh importance, especially in view of the need for petroleum from the Middle East. A fragment of the new proposition is stated by Robert Engler in *The Brotherhood of Oil*: "Arms to ensure a supportive energy-flow in turn require guaranteed energy sources. U.S. 'defense'—arms production and the armed forces—now account for at least 10 percent of national energy consumption. And the role of the United States as arms merchant and drill sergeant...is defended as necessary for balancing the flow of payments for foreign petroleum."

This provides the basis for analyzing a major dimension of the "hard-path" energy strategy assembled in the past few years by the corporate elite. Lest the reader ask, for example, why the U.S. mightn't cease its "flow of payments" to the Middle East by substituting solar energy for imported oil, the answer is that American corporations would much prefer to recoup those oil dollars through increased trade with the Arab nations. Thus, in a study for NSF by a Washington-based think-tank called Resources for the Future, the author asserts that although "imports of foreign oil will cost the country over $30 billion a year, the net effect on the U.S. balance of payments would be much less. The oil exporting companies will want to spend at least part of their foreign exchange earnings on American products." In addition to direct purchases, the Arab nations will be looking for investment opportunities, and "the United States with the world's largest and best developed capital market should receive a substantial portion of these funds." The author unabashedly concludes that "some government intervention may be needed to assure that energy R&D adequately emphasizes solutions to the energy problems threatening trade."

In other words, the government is expected to restrain the development of technologies like solar power until the international oil profits have been distributed. This has the happy effect of shoring up the defense corporations, because, as Engler points out, so long as we are willingly dependent on oil from the Middle East, we must be prepared to wage war to protect our access to that oil. Enter James Schlesinger, former hawkish Secretary of Defense who clearly has not recovered from the humiliation of Vietnam and seems almost to relish the prospect of a military contest with the Soviet Union.

Schlesinger was at his indignant best in a speech he gave at the Pentagon in 1975 on the occasion of his dismissal as Secretary of Defense. Referring both to what he views as a crippled U.S. defense posture and to U.S. dependence on OPEC oil, he said, "The underlying reality is that at no point since the 1930's has the Western world faced so formidable a threat to its survival...The growth of economic interdependence, notably in energy supply, implies that the industrialized world cannot survive without imports, massive in volume, from the less developed nations. These nations are no longer under Western political control and are exhibiting increasing hostility to the Western world and Western concepts of governance."

Schlesinger invoked the spectre of the Soviet Union, warning his audience that "the Soviets have been upgrading their airlift capabilities as part of a dramatic improvement of their mobility forces, which in the future will be able to intervene...in areas such as the Middle East." He implored the generals not only to beef up the American fighting machine but to accept the "historic destiny" of the United States as guardian of the western world: "How America responds to its historic destiny will determine the shape of the international community in the last quarter of the twentieth century...Without enduring American strength, Western civilization will not survive."

It is perhaps no accident that the Washington headquarters of the Department of Energy is the sprawling Forrestal Building constructed originally and not long ago for the Department of Defense. (This required a herculean evacuation effort by thousands of disgruntled Defense employees.) A recent article in *Nation's Business* suggests that the "threat of a new war in the Middle East, where an increasing amount of U.S. oil comes from, gave impetus to the move to set up the new department in the first place." The article does not mention that Schlesinger's deputy in establishing the Department of Energy was Thomas E. Reed, Secretary of the Air Force. It does declare that "in importance to the business community, the new Energy Department ranks near the top."

Another important link between U.S. energy policy and foreign policy is nuclear power. The May 6, 1977 issue of *Science* magazine carried an article on a Ford Foundation study which heavily influenced major provisions of Carter's "National Energy Plan." That study, according to *Science,* "leaves no doubt but what a substantial expansion of both coal-fired and nuclear electrical generation will be necessary if dependence on insecure foreign sources of petroleum is to be reduced." This represents a crucial plank in the corporate energy platform: while we don't mind importing a certain amount of OPEC oil—the "world market" price of which will enable us to inflate the price of our own oil—we don't want to be so dependent on it that we couldn't fight a war without it. Nor do we want to import so much that it wrecks our balance of payments. We are advised, therefore, to level off our long-term demand for petroleum—while not of course hindering economic growth requiring more fuel—by installing a massive grid of coal and nuclear generating plants. We are further advised to establish an emergency reserve of one billion barrels of preferably foreign oil (which reserve is currently being pumped into U.S. salt domes).

Equally crucial to energy planners is the question of U.S. membership in an international nuclear family which the U.S. has nurtured for almost two decades. Industrial nations like Japan, West Germany, Spain, France, and Britain all have sought to reduce their dependence on imported oil through construction of nuclear reactors, thus providing a generous foreign market for sales of hardware, licenses, expertise, and uranium by U.S. corporations. Since most of these countries have committed themselves to the long-term allure of plutonium fast-breeders, they were understandably exercised at Carter's announcement in early 1977 that he was curtailing the U.S. breeder program while he pondered the wisdom of shipping plutonium all over the globe. It wasn't long before Schlesinger appeared in Brussels, at a meeting of the Common Market, to deliver a little lesson in distinguishing between what Carter *says* and what he *does*.

"Energy," said Schlesinger, "as we see it jointly, is the main problem in relation to the revival of economic performance around the world." It followed that the Common Market nations certainly could rely on a continued flow of nuclear technology and fuels from the United States. Schlesinger implied that the only real snag in the U.S. nuclear program, aside from a few technical burrs like radioactive waste disposal, was a clutch of misguided environmentalists who were putting intense pressure on the Carter Administration. These doubters, said Schlesinger, "will never be convinced by arguments." But their effectiveness is sure to be fleeting: when the American people are faced "with the ultimate choice of alternative supplies of energy, the vast majority will accept nuclear power."

These examples of interface between U.S. energy policy and other priorities of the corporate state are intended primarily to show the seriousness with which the strategists have been obliged to view their task of forging a new energy plan. Given the complex of ills besetting them after Vietnam, it was rather inconvenient that solar energy arose not only as a competitor with coal and nuclear power but indeed as the core of a "soft path" social vision which could be a threat to the corporate order itself. Such a vision had no place in the new energy scheme of things which began to take shape in the closing months of Richard Nixon's first term in office as president.

II

IN SEARCH OF A CORPORATE
SOLAR STRATEGY: SMALL IS NOT BEAUTIFUL

"The rise of the global corporation," write Richard Barnet and Ronald Müller in *Global Reach*, "has been sparked by two fundamental tenets of the modern business faith: the cult of bigness and the science of centralization." The authors support their thesis with a quote from Carl Gerstacker, chairman of Dow Chemical Corporation: "It has been abundantly proved that size is often a gigantic advantage, and often, for certain tasks, a necessity. The problems of our times will require greater, bigger organizations than we now have, rather than smaller ones, for their solution...We must cast aside our outmoded notions of size and our fear of bigness." The successful application of these tenets of "bigness" and "centralization"— which are indispensable to the survival of the corporate American state in the last quarter of the 20th century—has required a few adjustments in the structural relationship between "government" and "business." It has required, indeed, that government more than ever reduce itself to an instrument at the disposal of the corporations in pursuit of their dream of "the Global Shopping Center."

23

Since the corporation perceives itself as "the primary engine of development," write Barnet and Müller, it follows that "the primary function of government is to enable the corporation to fulfill its promise. In the corporate vision of 1990, government no longer plays its traditional role under the Constitution." Instead of representing an "expression of national consensus," as determined through the democratic process, it is now "the real job of government...to perform certain services essential to the development of a good business climate. Most important," say the authors, "is the management of the economy. It is the function of government to stabilize the economy—to stimulate or to cool the economy as needed whenever the natural regulating mechanisms of the market fail to work (which seems to be much of the time)." Among the other key functions of government, "always in partnership with business," are to "control interest rates and other credit policy... impose price controls and wage controls," and "to act as a pump-primer of capital into the economy, through either the military budget or social programs. (At present the Department of Defense finances over one-half of all research and development costs in the U.S., including 90 percent of that done in the aviation and space industries.)"

When necessary, moreover, it is the lot of the federal government to rescue corporations which find themselves in trouble. "In 1969," write Barnet and Müller, "Lockheed, then the number one defense contractor, borrowed $400 million from a consortium of 24 banks. In 1970, while working on the L-1011 Airbus, the company ran into such serious financial difficulty that bankruptcy seemed imminent. Six representatives of the major banks that had lent the $400 million met in March of that year with Deputy Secretary of Defense David Packard to negotiate their rescue. The result was a $250-million loan guarantee, which the Nixon Administration proposed and the Congress narrowly passed in August 1971, by which the U.S. taxpayer relieved the rescuing banks of all risk." The Lockheed loan, write Barnet and Müller, "illustrates the hold that huge corporations—particularly those in transportation, utilities, and the defense industry—have on the rest of the society. Despite its inefficiency and mismanagement, Lockheed was subsidized by the U.S. taxpayer because of its very size."

As mentioned above, the hand in the glove of "bigness" is "centralization." Barnet and Müller contend that only "a centralized government, in which the president enjoys great discretion to employ one policy or another, can play such a delicate balancing role. Thus an important part of the new managerial vision of the new America is a stronger centralized executive and a relatively weaker Congress." It is implicit in the process of government centralization, which the major corporations have supported since the 1930's, that an ever closer intimacy be cultivated between the men who operate the government and those who run the global corporations themselves.

Indeed, write Barnet and Müller, "it has seemed quite reasonable to the last five (presidential) administrations to staff those parts of the Federal Government that regulate the economy with men on loan from the great corporations and banks." After 50-plus years of such a policy, the "Federal Government is (now) in a position to exercise little countervailing power against big corporations in large measure because of government-business interlocks in the most strategic areas of the economy." Worse still, with regard to the government's traditional planning function, "government planners do not have enough knowledge about the activities of global corporations to make the crucial planning decisions for the society. Thus the managers of the corporations have become the principal planners for the society by default."

These depressing observations might well have served as a field manual for the corporate-government executives who assumed the task in late 1971 and early 1972 of devising a new energy policy for the United States—including a program of solar development. It is the purpose of the remainder of this chapter to show that the only interests which were seriously considered in the evolution of that program were the interests of the corporations and government institutions whose executives dominated the planning operation itself.

Here, briefly stated, are the major principles in the corporate-government approach to solar energy. While it is doubtful that they existed in laundry-list form at the outset, they nonetheless emerge again and again throughout the process of developing a solar program. First: control the pace at which solar power becomes a viable force in the energy market, allowing time to maximize profits from fossil fuels and to consolidate the expanded electrical grid based on coal and nuclear power. Second: emphasize those applications of solar energy which are most compatible with the present system of capital-intensive, centralized power facilities, meaning primarily an emphasis on solar-electric concepts such as the power-tower, photovoltaic cells, ocean thermal generating plants, and giant expensive windmills (wind being an indirect form of solar energy). Third: minimize corporate risk in the evolution both of solar technologies and market opportunities, which suggests a program of heavy government subsidies and a willingness to let small companies assume as many of the early market hazards as possible. Fourth: mesh the burgeoning solar market into the larger corporate market by absorbing small successful firms (or their ideas), emphasizing mass production, and placing distribution under the control of utility companies. Fifth: deter the public from identifying solar energy as a possible means of altering the present economic and geopolitical structure of the United States (small is not beautiful). A sixth principle, by which the other five are made possible, is one which governs the more general corporate energy strategy as well: centralize control over U.S. energy decisions in a single, easy-to-reach place, e.g. a cabinet-level department.

There have of course for years been varying degrees of centralized corporate control over energy decisions, mainly in the form of executive access to pivotal government officials, including the president. This was especially true during the Nixon/Ford administration. As a harbinger of things to come, Nixon's attorney general in 1971 granted written permission to a coterie of major oil companies to engage in secret "collective bargaining" with the OPEC nations to try and impose order on what by then had become a chaotic situation of arbitrary price increases by various OPEC member states.

In assuming such a role, the oil companies were not only in violation of federal antitrust laws, they were actually performing a function normally reserved for government itself. "The companies," writes Robert Engler, "met at Teheran and Tripoli to negotiate with OPEC. They maintained a London Policy Group, composed of senior corporate executives, to establish the terms of reference for the negotiating teams. There was also a New York group and various specialized back-up committees." The result of these meetings was a corporate validation of the OPEC cartel and a forthcoming schedule of oil-price inflations that would finally explode in the Arab boycott of October, 1973—enriching the oil companies while wreaking havoc on American consumers. And yet, says Engler, "no public officials attended these meetings."

It was in 1972, however, that the first clear moves were made toward a broader and more deliberate corporate energy strategy that would, among other things, crimp the potential of solar power. One development was the publication of *U.S. Energy Outlook*, a $6-million study of America's energy future by the quasi-official National Petroleum Council, an advisory group dominated by energy companies and funded by the U.S. treasury. This document, not surprisingly, called for deregulation of oil and gas prices, exploitation of off-shore reserves and shale deposits, pilot plants for synthesizing fuels from coal, and expanded production of coal and uranium. It did not call for solar power. It did suggest that no one tamper with the existing structure of U.S. energy production and distribution.

Another development in 1972 was the launching of a $4-million Energy Policy Project by the Ford Foundation, which in the past has furnished the U.S. management elite with some of its more subtle approaches to difficult social and economic problems. (The president of the Ford Foundation is McGeorge Bundy, whom many will recall as a prime consultant to Kennedy and Johnson in their execution of the criminal war in Vietnam.) Directed by David Freeman, who now heads the Tennessee Valley Authority for the Carter administration, the Energy Policy Project eventually produced a voluminous series of studies and recommendations which became the backbone of Carter's NEP. In addition, Project staff members seemed to turn up at almost every important high-level energy dialog between 1972 and 1977. They were especially evident during a critical 1973 study by the Atomic Energy Commission to be discussed presently.

The most recent contribution by the Ford Foundation was a one-year study of *Nuclear Power, Issues and Choices,* mentioned above, which provided Carter with his ambivalent policy on the breeder reactor. Senior executive participants in this effort, chaired by Spurgeon Keeney of the MITRE Corporation, included Harold Brown (former president of Cal Tech, now Carter's Secretary of Defense), John Sawhill (FEA administrator under Nixon/Ford), Richard Garwin of IBM, W. Donham Crawford of Edison Electric Institute (EEI), and Hans Landsberg of Resources for the Future (RFF)—itself funded for the last 25 years by the Ford Foundation. Every one of these individuals and institutions, particularly MITRE, EEI, and RFF, are reappearing principals in the seven-year history of what now passes for U.S. solar energy policy.

A more explicitly solar-related development in 1972, which is detailed in a chapter to follow, was a project by the Southern California Gas Company to "demonstrate" the feasibility of "solar-assisted gas energy" systems installed in selected buildings in the gas company's service area—a pioneering step toward utility involvement in the solar heating and cooling industry. But far and away the most telling occasion of 1972 with a direct bearing on the future of solar energy was a study published in December of that year under the joint auspices of the National Science Foundation and the National Aeronautics and Space Administration. Entitled *Solar Energy as a National Energy Resource,* it appears to have been the first systematic attempt to locate solar energy in the proper corporate perspective. It was, to begin with, strongly influenced by an earlier technical study compiled by Associated Universities, Inc. for Nixon's Office of Science and Technology. The head of Associated Universities (a research consortium appended to Oak Ridge National Laboratory) was at that time Gerald F. Tape, a member of the General Advisory Committee for the

Atomic Energy Commission, which in 1975 became the General Advisory Committee for ERDA.

Chairman of the executive committee for the NSF/NASA "Solar Energy Panel" was Dr. Paul Donovan, director of NSF who would later be appointed Nixon's science adviser. Donovan's executive secretary on the committee was William R. Cherry of NASA, who subsequently became a branch chief in the Solar Division of ERDA. Two other members of the executive committee, Frederick Morse and Lloyd Herwig, the latter of NSF, also assumed executive positions in the Solar Division when ERDA was created two years hence. The "industrial consultant" to the executive committee was Paul Rappaport of the David Sarnoff Research Center of RCA, who was destined to rise in 1977 to the number one slot in DOE's new Solar Energy Research Institute (SERI) at Golden, Colorado. (RCA was an early participant in the development of solar-powered photovoltaic cells for the U.S. space program. In 1976 alone, the firm received nearly $1.5 million from ERDA to pursue its solar cell technology.) Still another member of the executive committee was Milton Searl, president of the aforementioned Resources for the Future.

It is significant that in addition to those named, a majority of the other panel members were employees either of NSF, NASA or universities and corporations with a record of heavy involvement in the aerospace programs which were under assault in a post-Vietnam backlash sweeping Congress at about the time the solar panel was conducting its sessions. Among the corporations represented were IBM, AT&T (Bell Laboratories, which for two decades has administered DOE's Sandia National Laboratory, site of much of the nation's most expensive solar research), TRW, United Aircraft, Monsanto Chemical, Dupont, and the Southern California Edison Company (currently under contract with DOE to operate a $130-million solar-thermal electric generating plant in southern California). The banking community was represented by P. Richard Rittleman of the Mellon Bank in Pittsburgh.

Among the universities on the solar panel were Harvard (Business School), Carnegie Mellon University, University of

Wisconsin, UCLA, and Cal Tech, which manages DOE's Jet Propulsion Laboratory, formerly an adjunct of NASA. The University of Delaware (a Dupont protectorate) sent a professor named Karl Boer, head of the university's Institute for Energy Conversion, who for several years had been funded by NSF and NASA ($1 million through 1974) to conduct research on a cadmium sulfide photovoltaic process. A portion of those funds helped pay for the construction in 1973 of a showcase solar house called Solar One. It was also in 1973 that Dr. Boer formed Solar Energy Systems (SES), a private company through which he intended to market his photovoltaic cells. Later that year, Dr. Boer transferred the majority of SES stock to Shell Oil Company, controlled by Mellon and Dupont interests. Dr. Boer has since divided his time between SES and his position at the University of Delaware. In 1976, ERDA granted the Institute for Energy Conversion nearly $2 million to continue development of the cadmium sulfide process. In 1977, Dr. Boer was awarded an additional $480,000 to spiff up his solar house and to study ways of developing a wider commercial market for photovoltaic cells.

Another school on the solar panel was Colorado State University, represented by Dr. George Löf, who is now president of Solaron, Inc., a major manufacturer of flat plate solar collectors and accessories. Colorado State in 1975 was awarded an NSF grant of $130,484 to perform an "analysis of solar thermal electric systems." In 1976, Dr. Löf's Solaron company was granted a total of $157,285 in ERDA funds, disbursed through HUD, to subsidize the installation of Solaron systems in selected residential units. The firm received $918,000 for the same purpose in 1977 and 1978. Listed recently on the New York Stock Exchange, Solaron closed a deal in August of 1977 with the Enersol subsidiary of Southern Union, Inc.—whose other properties include the Southern Union Gas utility in central Texas—to handle distribution of Solaron systems in the southwestern United States. In addition to his duties at Solaron and Colorado State, Dr. Löf is a prime consultant both to DOE and to Resources for the Future, as well as an officer of the Solar Energy Industries Association (SEIA), a trade organization financed and controlled by a core of *Fortune 500* corporations which have recently entered the solar field. (The curious character and behavior of SEIA will be discussed in Chapter XI.)

Sitting with the universities and corporations on the solar panel was an organization which has long bridged them both: the Arthur D. Little Company of Cambridge, Massachusetts, established in 1886 to perform "contract" research for industry and government. The representative for A.D. Little was Dr. Peter E. Glaser, a company vice-president who in 1968 conceived the idea of beaming solar energy to the surface of the earth by means of giant orbiting space satellites equipped with photovoltaic cells and microwave transmission devices. In 1971, Dr. Glaser assembled a consortium of three corporations— Grumman, Raytheon, and Spectrolab (a subsidiary of Hughes Aircraft)—to incubate the concept, and in 1972-73 the consortium received $200,000 from NASA to perform an "engineering feasibility study." The study was deemed positive, and the consortium is now proceeding with a three-phase plan for putting into orbit by the year 2000 a satellite expected to weigh 20,000 tons with a surface area of 44 square kilometers. In addition to this research, which has cost the government nearly $10 million since 1971, the A.D. Little Company was funded in 1974 by a group of 80 U.S. and foreign corporations to conduct a broad survey of "solar business opportunities."

Similar observations could be made about most of the other participants on the 1972 NSF/NASA solar panel. It thus afforded a preview of every important feature of the corporate-government strategy still to be fully evolved: (1) centralized control through the science adviser to a president uncommonly amenable to the corporate viewpoint; (2) domination by government agencies, universities, corporations, and utilities biased toward high-technology research and development, most of whom, indeed, had direct financial and professional interests in finding (or creating) a successor to the shrinking aerospace program; (3) a "revolving door" between corporate and government executive positions; (4) a consensus on the need for government subsidies to minimize corporate investment risk while nonetheless aiming at a profitable market for new technol-

MANY OF THESE MEN SPLIT FROM THEIR CO's TO START UP NEW SE FIRMS

ogies; (5) utilization of seemingly neutral universities as conduits for public money expended on behalf of corporate projects.

The programmatic and fiscal recommendations of the solar panel—which set a precedent still observed by energy bureaucrats—were scarcely surprising. "Before solar energy becomes a major source of clean energy for our nation," wrote the panelists, "it will require the involvement of industrial ingenuity and productive know-how to produce economic hardware and services." Most U.S. companies, however, "are looking for short term projects for their new enterprises" because "long range projects present great risk, and investment capital is scarce unless there is a high probability of return in a major line in their business." It follows that "the Government must make a long range commitment" and "the incentives must be substantial so that future profit is assured...that makes industrial investment pay off."

Then came the nitty-gritty of dividing the projected spoils among the various technologies to be pursued. The panel had calculated that by the year 2000, solar energy could provide: 10 percent of the energy required for the heating and cooling of America's buildings; 10 percent of the "gaseous fuels" (methane and hydrogen); one percent of the "liquid fuels" (alcohol); and one percent of the electrical power. As flawed as these projections are, they would still seem to indicate a lion's share of the R&D budget for solar heating and cooling technologies, which happen to be precisely those most conducive to small-scale applications, and a hefty share for "gaseous fuels" as well. Instead, for the heating and cooling of buildings, the panel recommended a 15-year budget of $100 million, $370 million for "gaseous fuels," and $3.5 *billion* for solar electric technologies (including the power-tower, photovoltaics, wind energy conversion, and ocean thermal). So much for Round One.

Two further corporate developments in 1972 bear mentioning here. One was a pronounced trend, seemingly more than coincidental, toward sudden corporate interest in the possibilities of solar technology both as a source of federal R&D money and as a long-term commercial option. In an 18-month

period from 1972 to 1973, according to a survey commissioned by ERDA, the following companies took action with regard to solar: Alcoa, Corning Glass, General Electric, Grumman, Honeywell, Owens-Illinois, PPG, Revere Copper and Brass, Enthone, Inc. (a subsidiary of the giant ASARCO conglomerate which also controls Revere Copper), Texas Instruments, TRW, Westinghouse, Bell Labs, COMSAT (a federally chartered communications satellite operation), Dow-Corning (jointly formed by Dow Chemical and Corning Glass to manufacture silicone, a key material in solar cells), IBM, Arthur D. Little, Mobil Oil, Motorola, Shell Oil, Exxon (which acquired two solar subsidiaries), Hughes Aircraft, General Atomic (jointly owned by Gulf Oil and Royal Dutch/Shell), McDonnell Douglas, Martin Marietta, and Boeing. It is important to remember that these corporate gestures occurred prior to the Arab oil boycott in October, 1973, hence prior to the national clamor for "energy independence" and alternatives.

A possibly related and certainly more ominous development in 1972 was the quiet christening of a new elite group of corporate executives called the Business Roundtable. The group was formed, under the leadership of Dupont chairman Irving Shapiro, from two smaller groups: the Labor Law Study Committee and the Construction Users Anti-inflation Roundtable (nicknamed "Roger's Roundtable" after Roger Blough of U.S. Steel). Of this group Harry C. Boyte has written for *In These Times* that "the continuing problems of the American economy in the 1970's have evaporated what used to be known as corporate liberalism and have produced instead a savage politics of self-interest and plunder not overtly advocated for decades. The Roundtable has been a major architect of such a politics."

According to Boyte, the group has worked both overtly and covertly to sabotage major legislation on behalf of consumers, environmental concerns, workplace safety, and full employment. It has "used a variety of forums to propound the thesis of a 'capital shortage' facing private industry that would require huge increases in profit levels in the coming decade." It has thus been at the barricades, for example, in the campaign to deregulate prices of oil and natural gas. So powerful is the group that in mid-1977, when it called for "less government regulation and (more) government incentives to spur corporate investment," Vice President Mondale appeared before its annual meeting to pledge "that the (Carter) administration's 'central role' would be to help stimulate such investment." The existence of the Business Roundtable has a direct relevance to solar energy policy which will be discussed in a later chapter.

While the NSF/NASA panel of 1972 laid the groundwork for a national solar energy policy, it was a series of events in 1973 and 1974 which sealed the fate of solar as a bit player on the energy stage. In June, 1973, President Nixon instructed Dixie Lee Ray, head of the Atomic Energy Commission, to produce an inventory of U.S. energy R&D requirements through 1979, along with guidelines for disposing of a projected federal R&D budget of $10 billion. He also appointed an Energy Research and Development Advisory Council whose mission, according to Robert Engler, was "to provide independent review of existing and proposed national energy programs and recommendations for new programs to the White House's Energy Policy Office." This group drew its membership "from the ranks of research vice-presidents of such corporations as Exxon, Consolidation Coal (Conoco), Consolidated Edison, General Motors, and IBM." Chairman of the group was Paul Donovan, Nixon's science adviser.

Engler writes that in November of 1973, with the nation stunned by the OPEC boycott then in force, this corporate advisory group was "meeting in as much secrecy as it could muster. There was the now familiar one-day public announcement for its meetings," and "a discussion of the president's proposed $10-billion energy research and development program was closed for 'national security reasons,' in violation of the Federal Advisory Committee Act of 1972." Other subjects of group discussion "included ways to meet the need 'to maximize opportunities for cooperative effort between industry and government,' and 'to consider ways of "pulling" technological

SECRECY
IN MEETINGS

THIS IS BEGINNING
TO SOUND LIKE AN
INVESTIGATION

Corporate Solar Strategy   35

innovation in energy system through adjustments in regulatory and marketplace incentives.'" There was in the group, says Engler, "an evident bias toward fossil fuels and uranium." At one of their meetings, "the council chortled over a scientist's presentation regarding the potential of solar power, wind power, and sewage conversion."

Government officials who addressed the group included John C. Sawhill (later to become FEA administrator, now a board member of Consolidated Edison and president of New York University), who, according to Engler, "discussed anti-trust and patent laws: 'We really want to get industry's view of patent policy...We can probably administer patent law in a flexible way.'" Sawhill derogated the liberal idea of government research corporations in which "'you lose the advantage of the type of control we want to exercise in COMSAT-type corporations—which let industry in on the decision-making.'" Sawhill further remarked that "'we are very anxious to get uranium enrichment [the one stage then not under industry control] into the private sector.'" (Sawhill is a member of David Rockefeller's Trilateral Commission, founded in 1973 as part of the corporate crusade to salvage international capitalism. Some observers believe that the Trilateral Commission, under the tactical direction of Zbigniew Brzezinski, engineered the election of Commission-member Jimmy Carter to the presidency in 1976. Sawhill, at any rate, co-authored a 1978 Commission study on U.S. nuclear policy to be discussed in Chapter VIII.)

Meanwhile, another corporate advisory group which Nixon had convened—the Emergency Petroleum Supply Committee (EPSC), also dominated by energy company executives—was conducting secret meetings in liaison with the National Petroleum Council. (A new and powerful member of the latter group was Hollis M. Dole, former assistant secretary of the Department of the Interior who had left his post to work for the Atlantic Richfield Company [ARCO]. Dole still works for ARCO—in its Washington office—and recently was appointed to an important DOE advisory committee. ARCO, moreover, had by mid-1978 bought and merged its way deeply into the new solar energy market. The meetings of this group, too, writes

Engler, were closed for reasons of "national security" and alleged discussions of "trade secrets."

Such rationales were even used "to provide justification for closing the doors to a meeting held in Exxon's headquarters in New York." Engler adds that a "simpler solution...was to fly to the Bahamas. Some fifty-five corporate energy executives and eleven government officials including Secretary of the Treasury George Shultz, William Simon, and White House energy adviser Charles DiBona (now head of the American Petroleum Institute) did that five days before President Nixon's energy message in November 1973 in which he...offered the energy industry just about everything it had sought, from opening up the naval (oil and oil shale) reserves and temporary licensing of nuclear plants without public hearings, to modest strip mining standards and the suspension of natural gas price regulation." (The above mentioned George Shultz is now vice-president of the Bechtel Corporation, a major DOE contractor for solar, nuclear, and fossil fuel technologies. Shultz is also a ranking member of the Business Roundtable and of Carter's Labor-Management Group, to be examined more closely in Chapter III.)

While the energy companies were doing their thing in the Bahamas and Dixie Lee Ray was doing her thing in Washington, America's electric utilities were likewise preparing an offensive that would ultimately impact on solar energy policy. The national lobbying trust for the utilities is the Edison Electric Institute (EEI), headed until 1978 by W. Donham Crawford, and it was in 1973 that EEI abandoned its obsolete research arm, the Electric Research Council, in favor of a new and more aggressive organization called the Electric Power Research Institute (EPRI).

The utility industry, like other components of the corporate sector, perceived itself in 1973 to be in potentially serious trouble. Rising fuel costs, general price inflation, and a shortage of investment capital, not to mention delays in the nuclear electrification program, were imposing unacceptable strains on the utilities. They needed a stronger utility brotherhood—to hold the line against such pressures as mounting public resistance to rate increases and nuclear power plants—as well as a souped-up research apparatus to try and lower costs through technological refinements. They sought these objectives through the creation of EPRI, headquartered in Palo Alto, California with a branch in Washington, D.C., and they sent it on its way with a first-year budget of $50-million of the ratepayers' money.

It is a measure of EPRI's importance to the corporate elite in general that the new organization was staffed by executives "borrowed" from numerous other sources. The founding president of EPRI was Chauncey Starr, who relinquished a position as dean of the School of Engineering and Applied Science at UCLA. Prior to that, he had served as a vice-president at Rockwell International, an aerospace and nuclear concern heavily endowed by DOE, and as president of Rockwell's Atomics International Division. EPRI's director of Fossil Fuels and Advanced Systems (including solar energy) is Richard Balzhiser, who came to the organization in June, 1973 from the White House Office of Science and Technology. There, according to an EPRI publication, "he played a primary role in energy R&D and policy planning." (Balzhiser later served on a relevant ERDA advisory committee.)

Another EPRI division manager was lifted from the Commonwealth Edison Company of Chicago, while still another is "on loan" from General Electric. Resources for the Future pitched in by loaning EPRI a Senior Fellow named Sam H. Schurr, who headed the organization's Energy Systems, Environment, and Conservation Division for two years before returning to the fold at RFF. (Schurr is now directing a two-year project for RFF entitled "U.S. Energy Strategy for the Future," funded by RFF/Ford [$200,000] and the Andrew W. Mellon Foundation [$500,000]). EPRI's intriguing relationship with DOE, the U.S. Congress, and a host of other institutions and corporations, both American and foreign, will be explored presently.

*****

On December 1, 1973, Dixie Lee Ray submitted to President Nixon the report he had ordered on U.S. energy R&D priorities. It was entitled *The Nation's Energy Future*. In her letter of transmittal she wrote that the study, "as requested," was performed "under the general guidance of the (White House) Energy Policy Office. It has also benefitted from the active participation of those Federal Agencies most concerned with energy research. Additionally, there has been widespread consultation with representatives of the private sector, including a broad range of energy industries." She must have received the Understatement of the Year Award for that one.

To direct and coordinate the actual work behind the study, Dr. Ray appointed an "Overview Panel" of senior government officials chaired by Stephen Wakefield, who had recently succeeded Hollis M. Dole as assistant secretary for energy and minerals of the Department of the Interior. Wakefield was a lawyer from a classy Texas firm with many clients in the oil business, and Engler describes him as having spoken on behalf of that business "with a purity of dedication to profit-making that might even have embarrassed oil executives. He hailed the integrated industry as the nation's principal asset for obtaining energy...Responsible for developing a national energy policy, Wakefield warned against any 'dismantling' of the integrated structure. This 'would so weaken the attractiveness of U.S. markets that this act in itself would assure that the United States stood at the end of the line of claimants in a world where energy scarcity is likely to be the normal condition for many years to come.'"

One of hardliner Wakefield's colleagues on the Overview Panel was Deputy Secretary of the Treasury William E. Simon, former Wall Street Stockbroker (Salomon Brothers) who would soon become FEA administrator and then Secretary of the Treasury. In an interview with *Science* magazine early the following year, during his tenure with the FEA, Simon said that a chief target of Nixon/Ford's "Project Independence" was "bringing on additional supplies (of fossil fuels) to bring the game plan forward. We commenced about a month ago the integration of the Federal Energy Office (FEO) into the economic mechanism of government, which, if you know government, is an extremely important step. Economic policy has traditionally been set by the Secretary of the Treasury, the head of OMB, and the head of the Council of Economic Advisers, Herb Stein. That was the troika, and I sat on the troika as Deputy Secretary of the Treasury...Now we have started another group to deal with energy. We have added FEO, the Council of Economic Policy, and the Chairman of the Cost of Living Council."

The latter official, head of the Cost of Living Council, was at that time a Harvard business economist named John Dunlop, whom Ford would appoint as Secretary of Labor. Simon told *Science* that "what we need most of all (to resolve the energy crisis) is a well-functioning market...And that is what John Dunlop and myself have been working on very diligently, to make sure that these promises of the future can be fulfilled by the capacity of American industry." The significance of this connection between Dunlop and energy policy, including his ties with the Business Roundtable, will be apparent momentarily.

Wakefield's Overview Panel was assisted in defining the parameters of the energy R&D study by a series of workshops at Cornell University whose participants constituted a virtual Who's Who of the U.S. energy establishment. Among them were Fred L. Holloway (an Exxon vice-president now a consultant to DOE), John O'Leary (director of licensing at AEC, now Schlesinger's deputy at DOE), Harry Perry of RFF, W. Kenneth Davis of the Bechtel Corporation, Chauncey Starr and Milton Levenson of EPRI, Thomas O. Paine (then chairman of General Electric), Leif Olsen of First National City Bank, John Corcoran and M.L. Sharrah of Conoco, Sol Buchsbaum of AT&T (Bell Labs), and Carl E. Bagge, president of the National Coal Association (now a member of a DOE advisory committee).

Once these gentlemen had agreed on the scope and direction of the study, they divided the production chores between 16 ostensibly separate subpanels defined according to particular goals and energy technologies. Each subpanel was composed of five to ten government officials with a larger phalanx of non-government "consultants." The subpanels included "Resource Assessment," "Mining—Coal and Shale," "Conversion Techniques," "Solar and Other Energy Sources," "Advanced Transportation Systems," "Systems Analysis," etc.

Any attempt to catalog the 282 "consultants" on these subpanels, along with their corporate and university affiliations, would be redundant. Suffice it to say that the corporate viewpoint was exceedingly well represented. General Electric, for example, had six executives on board. Exxon had seven. Gulf had four, Westinghouse two, United Aircraft six, Conoco five, Amoco four. Other corporations included Shell, Boeing, Dow, Chrysler, G.M., U.S. Steel, Alcoa, Mobil, ARCO, and Southern California Edison. University panelists came from UCLA, Wisconsin, Penn State, Maryland, MIT, Stanford, Rochester, Carnegie Mellon, and the University of Delaware. The MITRE Corporation, a research think-tank along the lines of Arthur D. Little Company, had five representatives on the subpanels, while EPRI and the Ford Foundation had seven each.

Of more immediate interest are the composition and fate of Subpanel IX—"Solar and Other Energy Sources." It was, of course, dominated by the same kinds of institutions as the NSF/NASA study of the year before, with a number of the same companies, universities, and individuals present. NSF had no fewer than 18 employees on the subpanel, including its chairman, Alfred J. Eggers (assistant to Paul Donovan and later an executive in the Solar Division of ERDA). NASA had seven. AEC was represented, as were Bell Labs, Grumman, Consolidated Edison (New York), the National Petroleum Council, Carnegie Mellon University, Johns Hopkins, Cal Tech (Jet Propulsion Laboratory), University of Maryland, Sandia Labs, MITRE, and the Ford Foundation (Energy Policy Project).

Among the individuals on the subpanel were at least four— William Cherry, Frederick Morse, Lloyd Herwig, and William Woodward—who had served on the executive committee of the 1972 NSF/NASA panel, three of whom later went to the Solar

Division of ERDA, as did still another Subpanel IX participant, Richard Bleiden of NSF. (Following his tenure as assistant director for solar electric applications at ERDA, Bleiden moved on to the vice-presidency of ARCO Solar, Inc., which has since become a significant contractor in the federal solar program.) Bell's representative on the subpanel, F. Smits, was also present in 1972. There were, as far as can be determined, no representatives either from small business or from the Small Business Administration. There were representatives from the Departments of the Army and Navy.

The fruit of the subpanel's labors was a menu of "FY 75-79 Program Objectives" for solar R&D which are stunningly close to the program which DOE has pursued to this day. Only one of six proposals referred to the heating and cooling of buildings. Four of the remaining five were directed toward solar-electric applications: (1) construct a 10-megawatt (MW) experimental power-tower (now under contract to Boeing, Martin Marietta, McDonnell Douglas, etc.); (2) construct a (huge) 100-killowatt experimental windmill pursuant to installing a 10MW "windmill farm" (varieties of which are under contract to Boeing, Lockheed, Grumman, etc.); (3) develop the ocean-thermal electric concept (TRW, Lockheed, Hughes Aircraft, Carnegie Mellon, etc.); (4) push photovoltaic cells to the mass production stage (RCA, Shell, Exxon, Mobil, ARCO, etc.) The sixth proposal related to experiments with biomass energy production (fuels from garbage and organic materials), which ERDA has since consigned to the Bechtel Corporation, Battelle Institute, Cal Tech, Cities Service, G.E., MITRE, etc.

To fund this R&D, the subpanel recommended two different five-year budgets, one for an "orderly" program at $1 billion, the other for a "minimum" program at $400 million. The catch was that each of the subpanel's recommendations had to pass before Stephen Wakefield's Overview Panel, and he only had $10 billion to spend. So he and his colleagues devised a point-grading system for assessing the relative value—presumably to the nation—of each of the competing technologies. Their criteria included "Adequacy of Scientific Base," "Feasible Absorbable Investment," "Cost of Substitutes," "Environmental Acceptability,"

"Adequacy of Resource Reserves," "Savings in Petroleum," etc.

When the scores were tallied, "Coal and Shale Processing" ranked near the top with an "unweighted" total of 42 points, so it got a five-year R&D budget of $1.3 billion. "Nuclear Fission" (including the breeder reactor) won 39 points and a budget of $4.1 billion (either Wakefield or Dr. Ray evidently had the option of adding a few dollars here and there for good measure). "Advanced Transportation Systems" took 35 points and $505 million. "Solar" was last with an embarrassing 27 points and a slashed budget of $200 million (2 percent of the total, which isn't terribly below its "unweighted" rank in DOE's budgets for 1979 and 1980).

In addition to program and budget priorities, the Ray/Wakefield document offered suggestions regarding the "Federal Government's responsibilities in the national energy R&D program." It noted, first of all, that whenever "national goals coincide with those of private industry, then private industry should be encouraged to attain the national goals." The government's task is to "identify, in conjunction with private industry, the research and development needed to reach those goals." Whereupon the government should "maximize industry participation" and "tailor participation methods to individual industries," while expanding "Government facilities only when no capability exists nor can be created in the private sector." The government is further obliged to "develop Federal measures to reduce the commercial uncertainties of early application of new technologies," and, perhaps most important, "to press vigorously for the establishment of a single government organization (Energy Research and Development Administration) to coordinate the national program." In view of the fact that what the authors mean by "private industry" is Exxon, Boeing, and General Electric, it can be concluded that the Ray/Wakefield effort represented the loss of Round Two for small-scale solar energy.

Four months later, in April of 1974, a conference was held in Washington, D.C. under the auspices of Resources for the

Future with funds provided by NSF. Its theme was "U.S. Energy Policy: The Role of Economics," and its purpose was to solicit high-powered response to a treatise on the subject prepared in advance by economist John E. Tilton of Penn State University. In his preface to the conference proceedings, RFF co-director Hans Landsberg remarked that to some observers "all this (conference dialog) may seem an exercise in futility now that the country is firmly embarked upon a large energy R&D program and the time for thinking about how to fashion it is past."

The conference did seem rather an exercise, an academic picking of nits from the hairy flanks of Nixon's new $10-billion strategy for "Energy Independence." Despite a few squabbles over such concerns as "risk pooling and discounting," the conference validated most of the key provisions and underlying principles of the Ray/Wakefield plan. The conferees were unanimous, for example, on the desirability of a centralized federal energy mechanism, and there was no significant debate over the ranking of the various technologies to be capitalized, particularly the breeder reactor. Above all was a consensus on the *need* to develop new fuel resources: "Energy," agreed the participants, "is the engine or force that drives the economy." They contended that "political liberty, the emancipation of women, and other cherished features of Western civilization... would be impossible without abundant and low cost energy." Some of the participants compared energy with gold.

More fascinating than the record of the conference itself is the roster of persons who sacrificed the two days necessary to attend. John O'Leary was there from AEC, along with Exxon's Fred L. Holloway, Bechtel's W. Kenneth Davis, and Bruce Hannay of Bell Laboratories. EPRI dispatched Sam H. Schurr, while the gas utilities and corporations were represented by Henry R. Linden, president of the Gas Research Institute and a member of ERDA's powerful General Advisory Committee. Among the five RFF representatives (not counting Schurr) were Hans Landsberg and Harry Perry. They were accompanied by an old pal from NSF named Paul Craig, who, according to Landsberg, was soon to be transferred "into the (White House) Office of Energy and R&D Policy, and thus into the very center of the topic." Also present were executives from the White House Office of Management and Budget (OMB) and Federal Energy Office (FEO), as well as the MITRE Corporation and Battelle Memorial Institute. Academies included Harvard, Yale, MIT, Cal Tech, Cornell, and the Universities of Denver, Minnesota, and Texas.

Thus, by mid-1974, America's corporations had not only identified an explicit new strategy for the nation's energy future—quite without consulting the American people who would bankroll that future—they had canonized it with a $10-billion budget. When next they commenced to tighten up the administrative machinery for executing their "game-plan," they started at the top—by planting virtually in the White House an elite "advisory group" of corporation and labor leaders which represented a new plateau in what Barnet and Müller have called the "business-government partnership." That advisory group and the chain of command through which it has effected its policy decisions—much to the containment of rational solar energy development—are examined in the following two chapters.

III

FEDERAL ENERGY ADVISORY COMMITTEES: CORPORATE SENTRIES & UTILITY WATCHDOGS

> The federal government remains honeycombed by a network of energy advisory bodies composed of leaders of the major corporations and trade associations, along with the usual decoration of "independents." These advisers define the acceptable bounds of policy alternatives, police their implementation, and in effect become the makers of public policy. They thus undercut the legislative process and distort responsible administration.
>
> —Robert Engler, *The Brotherhood of Oil*

Much of what was said in the previous chapter with regard to "centralization" of the government decision-making process, which is shared in turn with the government's "partners" in the corporate sector, may be applied to this chapter as well. In fact, the same corporations, utilities, and universities which dominated the major solar policy initiatives of 1972 and 1973 will now be found in key positions on those federal "advisory committees" charged with the task—in 1974 and 1975—of translating the government's new energy policy into a body of concrete programs, budgets, agencies, and administrators. These were critical years for the solar program in particular, because the goals and priorities established at this point, in conjunction with the "style" of the federal agencies involved, would etch a pattern likely to influence U.S. solar energy development for the remainder of the 20th century.

The significance of this period to the corporations themselves was apparent in the creation of a super-elite "advisory" council in September, 1974—ostensibly under the wing of Treasury Secretary William E. Simon, the Wall Street financier who had earlier served as Nixon's "energy czar" and Deputy Secretary of the Treasury. According to its charter, this new "Labor-Management Committee" was supposed to study and make recommendations "with respect to policies that may be followed by labor, management, or the public which will promote free and responsible collective bargaining, industrial peace, sound wage and price policies, higher standards of living, increased productivity, and related manpower policies...which could contribute to the longer-run economic well-being of the Nation." But the truth is that the primary business of the Labor-Management Committee for at least the first year of its existence was energy policy, pure and simple, most of which was formulated in direct personal liaison not with the Treasury Department but with President Ford. (If this was an abuse of its charter, no one need ever know, because—in violation of the Federal Advisory Committee Act of 1972—the charter was not submitted to any of the congressional committees with jurisdiction over "labor-management" concerns. It was submitted only to the Library of Congress and the White House Office of Management and Budget.)

The membership of the Labor-Management Committee, according to a 1977 Senate study of "Federal Energy Advisory Committees," was divided between eight representatives of "management" and eight of "labor." All eight "management" representatives were "the chief executives of some of the largest U.S. corporations," while the "labor" representatives were presidents of major international unions, including George Meany of the AFL-CIO. Riding herd on the group was a "neutral coordinator" named John Dunlop, the Harvard economist whom Ford had appointed as Secretary of Labor.

This amalgam of "advisers," says the Senate report, stands "in a class apart from the usual advisory committee, having direct personal access to the President, who attended the meetings (in 1974 and 1975) and received advice from the group both orally and in written communication." Meetings of the group "are closed to public participation, making clear that this is a confidential advisory process between the elite of the business/industry/union world and the President. However, the energy questions on which recommendations were (and are) formulated have far-reaching consequences for many sectors of the public."

In December, 1974—two months after its founding—the committee tendered a report which "covered a wide range of issues, including expansion of domestic supplies, energy conservation in order to reduce imports, increasing sources of existing energy, particularly coal and nuclear power. Research and development acceleration was urged," as was "petroleum stockpiling." In 1975, "formal recommendations were made on 'National Energy Policy' and 'Electric Utilities.' These recommendations were usually reported in person to the President."

Among the committee's suggestions were "tax-credit adjustments (for utilities), pollution control write-offs, extension of the Price-Anderson Act to continue Federal nuclear indemnity coverage, stock regulation measures," and a presidential "task force of experts to discover impediments to completion of electric utility plants." These "experts," opined the committee, should act "to relieve" such impediments, "and it was urged that environmental restrictions on energy be stretched out." The Senate study adds that the "report on electric utilities...recommended measures to stimulate electric utility construction. It was finalized on May, 21, 1975, and sent to the President with a request that he make it public and endorse it. This he did in a statement of June 13, 1975." (The Labor-Management Committee—with a slightly altered title and official status—is alive and well in Washington in 1979. Its current membership will be discussed later in conjunction with an ERDA advisory committee.)

When ERDA was activated in January, 1975, it absorbed not only the Atomic Energy Commission but most of the Commission's advisory panels, including the potent General Advisory Committee. Two more advisory committees were bequeathed to ERDA by the Interior Department, and ERDA created three of its own, for a total of 12 in June, 1975. These advisory bodies, along with sister groups in the FEA and other federal energy bureaus, formed an awesome complement to the Labor-Management Committee in terms of domination by utility and corporation executives. The symbiotic nature of this arrangement was exemplified in June, 1975, when an FEA "Electric Utilities Advisory Committee" voted to endorse "the President's Labor-Management Committee's recommendations to facilitate electric utility construction." That FEA committee, according to the 1977 Senate report, was chaired by W. Donham Crawford, then president of Edison Electric Institute.

The Senate inquiry found other varieties of symbiosis as well. It found, for example, that federal energy advisory committees are interlocked with corporations both "vertically," where an Exxon vice-president sits on a "Fossil Energy Committee," and "horizontally," where that same vice-president sits on a "Procurement Policy Committee." The study reveals that in 1975, out of a total of 2143 "advisers" on 61 energy advisory committees, "232 individuals served on more than one committee. These 232 people held 529 individual memberships, accounting for nearly 25 percent of the total energy advisory committee membership." Some individuals sat on as many as five committees, including Henry R. Linden, the gas executive who enjoyed a valuable position on the General Advisory Committee of ERDA. According to the Senate report, "Energy Production and Distribution Companies" had 53 executives on more than one committee, while "Utilities and Utility Associations" had 46 employees on multiple committees.

Still, the denser matrix of federal advisory interlocks occurs through firms as opposed to individual executives: Exxon was represented on 15 separate advisory committees in 1975. Texaco held 12 memberships, while nine or more each were occupied by the MITRE Corporation, Electric Power Research Institute, Consolidated Edison, Southern California Edison, Commonwealth Edison, Mobil, Panhandle Eastern Transmission, El Paso Natural Gas, and the Institute of Gas Technology (now the Gas Research Institute). Other corporations represented on multiple committees included Dow (8), U.S. Steel (6), General Electric (5), and Dupont (5).

Except for "the President's Labor-Management" group, the single most important advisory body during the critical formative years of the federal solar energy program was ERDA's General Advisory Committee (GAC). It had evolved, says the Senate report, "from a central and influential role in the AEC to an equally key role for ERDA, with a broadened mandate and membership to match the wider focus of the new agency." The GAC had "been designated by the (ERDA) Administrator as an 'umbrella' for the other ERDA advisory committees. This (was) accomplished in part by having GAC members sit as ex-officio members of the other committees." (William R. Gould, for example, a vice-president of Southern California Edison Company, served not only on the GAC but in an "ex-officio" capacity on ERDA's Geothermal Energy Advisory Committee.) The Senate study adds that the GAC was "expected to concern itself with the entire range of issues that ERDA deals with, including the weapons program. The committee can obtain ERDA staff studies on request," and members "have access to all kinds of ERDA operations. It is the only advisory committee in ERDA to have its own permanent staff of several people."

Due to its "obvious central role" in the federal energy program, the Senate report concludes that "this committee even more than the others would be of more general (public) interest," implying an obligation to open and keep records of its meetings and other activities—as required by federal law. But the reality was that "no transcripts of meetings were kept, and only very brief minutes were made available following the meetings. Most meetings (were) partially closed, following the previous practice of the GAC (under the Atomic Energy Commission)... The committee has evidently been considered a forum for discussion of operations problems, and it has been felt that such problems could not be discussed as frankly during open portions of meetings." In order to finance these secret forums, taxpayers coughed up between $150,000 and $200,000 per year for committee expenses from 1975 through 1977. (The apparent successor to the GAC, established by Schlesinger in October, 1977, is the "Energy Research Advisory Board." Its relative weight among DOE advisory committees may perhaps be measured by its annual budget of $425,000 and full-time staff of five employees, making it four times richer and better staffed than any other DOE committee. Among the members held over from the GAC are Charles J. Hitch of Resources for the Future and James E. Connor of First Boston, Inc. Newer members include Edward David of Exxon Research and Engineering, Sol Buchsbaum of Bell Laboratories, and Oliver Boileau, president of Boeing Aerospace.)

As of July 26, 1977, all nine members of the GAC had also been designated the "Solar Working Group" within ERDA, and a cursory analysis of that group, along with its official consultants, offers both a close-up look at the internal machinery of the "government-business partnership" and a means of understanding the otherwise mysterious trajectory of the federal solar energy program. Two of the "consultants" to the "Solar Working Group" were Paul Rappaport, the RCA executive who now heads DOE's Solar Energy Research Institute, and Richard Balzhiser, director of EPRI's Fossil Fuels and Advanced Systems Division who worked for Nixon as an energy adviser. The other four "consultants" were tied to major universities and national laboratories (Berkeley, Wisconsin, Michigan), while the "Contractor" for the group was the Manager of Solar Energy at the Stanford Research Institute (SRI), which consumed $487,000 of ERDA solar money in 1977. (SRI will be discussed presently in connection with a stonewall job.)

Chairman at once of the "Solar Group" and the GAC was Charles J. Hitch, president of Resources for the Future, whose earlier career plateaus included 13 years with the RAND Corporation, a blue-ribbon corporate think-tank which provided enormous tactical and strategic input into the U.S. effort to win the war in Vietnam. Dr. Hitch was also an Assistant Secretary of Defense under Kennedy and Johnson, followed by an appointment first as vice-president and then as president of the University of California until July of 1975, when he resigned to accept the presidency of RFF. He is currently a trustee of the Asia Foundation, a member of Rockefeller's Trilateral Commission, and a director of the Aerospace Corporation—which in 1977 alone received $2,602,851 worth of contracts from the Solar Division of ERDA.

One of Dr. Hitch's colleagues on the GAC was William R. Gould, cited above, who is both a vice-president and a director of Southern California Edison Company, as well as eleven other corporations and corporate trusts, including the Breeder Reactor Corporation and the Electric Power Research Institute. He is also chairman of the Executive Advisory Board for Nuclear Power Policy of the Edison Electric Institute and a former chairman of the Atomic Industrial Forum, the lobbying front for the U.S. nuclear industry. Gas corporations were further represented on the GAC by the aforementioned Henry R. Linden, who, in addition to his job as president of the Gas Research Institute (the research arm of the American Gas Association), is a member of the American Petroleum Association, a former employee of Mobil, and a director of Southern Natural Resources, Inc., of Birmingham. Dr. Linden's memberships on federal energy advisory councils from 1963 to the present, including the White House Energy R&D Council from 1973 to 1975, would require several pages to catalog. He was accompanied on the GAC by Dr. Gerald F. Tape, president of Associated Universities, former director of the Atomic Industrial Forum, and current U.S. representative to the International Atomic Energy Agency.

The nation's banking elite had two positions on the GAC. One was occupied by Dr. James E. Connor, Assistant to the Chairman of First Boston, Inc., where, according to his resumé, he works "directly with officers of the company on matters involving Planning and Analysis, Government Relations, and Broad Corporate Affairs." He served as Secretary to the Cabinet and Staff Secretary under President Ford, providing "liaison between Cabinet Officers and the President," and from 1972 to 1974 was Director of Planning and Analysis for the Atomic Energy Commission, representing "the Commission and U.S. Government in negotiations with foreign governments and the U.S. private sector." (Ironically, Dr. Connor's dissertation at Columbia University was entitled *The Soviet State Bank: a Study in the Development of an Instrument of Control.* He is also the author of "Prospects for Nuclear Power" in *The National Energy Problem*, published in 1973 by the Academy of Political Science.)

The second financier on the GAC, appointed by President Carter, was a familiar name from the Nixon/Ford administration—Frank G. Zarb. He was America's very first "energy czar," a title bestowed by the media upon his appointment in 1974 as head of the Federal Energy Administration. Zarb served concurrently (until 1977) as Executive Director of the White House Energy Resources Council, in which position, according to an ERDA release, he was "the President's primary advisor on National energy policy and programs. "Prior to that he was Associate Director of the Office of Management and Budget, with funding authority over all federal energy programs, and Assistant Secretary of Labor from 1971 to 1972." In a June 9, 1977 interview with *Energy Daily,* not long after Zarb had left the government to return to work for Shearson, Hayden, Stone on Wall Street, he told the magazine that his new job kept him "on the road 60 percent of the time...working on international energy and energy-related projects that take him to the Middle and Far East, though he (wouldn't) say exactly what those projects (were)." He "lamented" that his "billion-barrel baby, the oil stockpile, (was) not yet in place, because the chances of another (Mideast) oil embargo...'or other scenario that could stop the flow of oil on the high seas are very, very high.'"

Zarb also predicted that despite Jimmy Carter's presidential campaign pledges, he would not only deregulate domestic oil and natural gas prices—a move announced by Carter on April 5, 1979—but would plow ahead with nuclear energy development. Indeed, Zarb was "cheery" about the development of the breeder reactor, declaring that "as 1985 draws closer, the breeder will start to look a lot more necessary." Zarb has since become a general partner in the power-house brokerage firm of Lazard Freres, where he was "brought in," according to an item in *Time* magazine (January 22, 1979), "to help Lazard expand as an adviser to foreign governments in arranging large financings. He will doubtless be bolstered in this endeavor by his new position as "energy consultant" both to the House Commerce Subcommittee on Energy, chaired by Rep. John Dingell, and to Governor Jay Hammond of Alaska.

The remaining three members of ERDA's General Advisory Committee, though not as eminently placed in the U.S. corporate structure as those above named, had nonetheless some interesting ties with that structure. Dr. Ruth Patrick of Philadelphia was apparently the token "environmentalist" on the GAC. Besides her position as an adjunct professor of botany at the University of Pennsylvania, Dr. Patrick was a member of the Advisory Council of EPRI, the Executive Advisory Committee of the Federal Power Commission, and the Boards of Directors of the Pennsylvania Power and Light Company and the Dupont Corporation. A companion of the GAC, Dr. Michael May, was Associate Director at Large of the Lawrence-Livermore National Laboratory (operated by the University of California). His resumé states that Dr. May's earlier work for the Lawrence Radiation Laboratory included "technical leadership for the groups which conceived and designed new nuclear explosives for military and civilian uses." Currently a trustee of the RAND Corporation, Dr. May was evidently the weapons specialist, among other functions, on the GAC.

The labor representative on the committee was Martin J. Ward, General President of the United Association of Journeymen and Apprentices of the Pipefitting Industry of the United States and Canada—in other words the plumbers' union. He was also a department vice-president and member of the Executive Council of the AFL-CIO, as well as a trustee of the American Institute of Free Labor Development. (The latter organization, created in the late 60's with the blessings of the U.S. Departments of State and Labor, has occasionally been linked with shoddy American missions on behalf of "free enterprise" in other countries, including the conspiracy against Chile's Marxist president Salvador Allende, which led to his downfall and death in 1973.) Ward was present at a February, 1977 meeting of the AFL-CIO Executive Council in Bal Harbour, Florida, where the unionists issued a statement in support of a plan by Nelson Rockefeller to earmark $100 billion in U.S. tax money for a "corporate energy development fund."

It is clear, said the union men in Bal Harbour, "that coal and nuclear power are the basic energy sources for the near future." This means that "facilities to enrich uranium should be expanded and the procedures for licensing of nuclear facilities should be expedited to eliminate costly and unnecessary delays." It means further that "continued development of the liquid metal fast breeder reactor program must be pursued—this is essential to the nation's long-term energy needs." The union executives also endorsed a centralized Department of Energy and "establishment of an oil stockpile." Given the implications of these suggestions, particularly Rockefeller's "$100-Billion Plan," one is forced to question the sincerity of the AFL-CIO's demand, repeated at the Bal Harbour meeting, for a "break-up of oil monopolies and prohibition of oil companies from owning competing sources of energy." (If the AFL-CIO favors small-scale solar energy development, it was not apparent in Bal Harbour.)

Martin Ward's intimacy with the corporate elite of the United States has by no means been restricted to his membership on ERDA's General Advisory Committee. He happens also to be one of the eight "labor" representatives on the "President's Labor Management Committee" discussed above. This puts Ward and the other "labor" leaders (including the presidents of the Teamsters, Seafarers, Clothing Workers, Steelworkers, AFL-CIO) in very posh company indeed. Reginald E. Jones, for example, whom President Carter named co-chairman of the group (with George Meany), is Chairman of the Board of General Electric. The other "management" representatives, all chairmen of their respective corporations, are: Irving S. Shapiro (Dupont), Walter B. Wriston (First National City Bank), Arthur M. Wood (Sears Roebuck), Rawleigh Warner Jr. (Mobil), Thomas A. Murphy

(General Motors), Edgar B. Speer (U.S. Steel), John T. Harper (Alcoa), and Stephen Bechtel Jr. (The Bechtel Group, whose president is George Shultz, former Secretary of the Treasury and himself a new addition to the Labor-Management Group).

It might perhaps be said in fairness to "populist" Jimmy Carter that while he hasn't abolished this yoke of VIP's—with its direct "labor" link to the key policy organ in the federal energy program—he has at least altered its status from an official "advisory committee" to an "informal" advisory *group*. Such a change, of course, by dispelling the appearance of outright conflict with federal law, may actually work to the group's advantage. Meanwhile, according to an article by Harry Boyte for *In These Times*, the function of the Labor-Management Group remains nominally the same: "to advise President Carter on key issues and to help restrain the 'wage-price' spiral that government considers a major cause of inflation." Likewise unchanged is the group's "neutral coordinator," former Secretary of Labor John Dunlop, who is back at Harvard teaching economics in the School of Business. Finally, for those who haven't made the connection, each corporate member of the Labor-Management Group is equally a member of the Business Roundtable, co-chaired by Dupont's Shapiro and G.E.'s Reginald Jones.

IV

THE FEDERAL SOLAR BUREAUCRACY: A TALE OF REVOLVING DOORS

It should now be clear why the launching of ERDA in January, 1975 did not produce even a modicum of the new opportunities for small-scale solar research and development which many had expected of the agency. Few but the wildest dreamers had hoped for a government program—in response to the "energy crisis"—which would move toward decentralization and rationalization of America's entire energy system. But thousands had hoped for government support of small solar firms and inventors, perhaps in tandem with environmental or community development programs. Even they had failed to account for the long-established weight of the Atomic Energy Commission and its melange of ties with the corporate-technological elite, all of which were transformed virtually undisturbed into the programmatic machinery of ERDA. Indeed, the histrionic flurry of congressional initiatives from mid to late 1974, sanctioning not only ERDA but the Solar Development Act and Senator Henry Jackson's $20-billion "non-nuclear energy research and development" porkbarrel, was a ram-rod victory for the corporations in Round Three of their struggle for still more control over America's energy future.

ERDA's inevitable bias was augured from the outset in the staffing of the new agency, as well as in its policies and procedures. The middle notches of the administrative totem-pole, including those in the Solar Division, were plugged with holdovers from the aerospace and nuclear operations of NASA, NSF, and AEC (partially discussed above). Higher positions were filled with imports both from other federal agencies and from interested corporations. ERDA's first administrator, for example, was Robert C. Seamans, Jr., a former RCA executive who had also been Secretary of the Air Force and Deputy Director of NASA. Seamans expressed a central and oft-cited component of ERDA's self-image when he wrote in *Public Utilities Fortnightly* (September, 1976) that the "unique feature of ERDA, compared to NASA, DOD, and other federal agencies, is that we do not use our end products. We do not have missions to go into orbit or to the moon. Our end product is the development and demonstration of technologies that are useful only when applied on a significant scale by private industry. The role of the private sector is paramount...It is the primary vehicle to resolve the nation's energy problems."

When Seamans left ERDA in January, 1977, he accepted a position at MIT (where he is now dean of the School of Engineering) and was promptly elected to the Board of Directors of Combustion Engineering, Inc., which does a lot of business in advanced fossil fuel and nuclear technologies both with DOE and with EPRI. James Schlesinger then appointed Robert W. Fri of Union Carbide, Inc., another major recipient of federal energy funds (over $1 billion in 1977) to be Acting Administrator of ERDA pending the establishment of DOE. Fri joined no fewer than eight executive counterparts from Union Carbide in the senior echelons of ERDA, according to a study by Common Cause. And within a year of his departure from the agency, according to an item in *Energy Daily* (January 12, 1979), Fri became president of Energy Transition Corporation, a brand new outfit concocted by a "group of prominent businessmen and former federal officials...in the hope that their combined finan-cial, political and technical skills can assist in the commercial introduction of new technology." (Among the company's directors, adds *Energy Daily,* "are Frank Zarb, once administrator of the FEA, now a general partner at Lazard Freres, and William Casey, once chairman of the Securities and Exchange Commission.")

The above-cited study by Common Cause—an analysis of interlocks between ranking government officials and corporations receiving government contracts—also found six former executives of General Electric in key ERDA positions, four from Westinghouse, three from United Aircraft, and one or more from such corporations as Western Electric (AT&T), General Atomic (Gulf Oil), RCA, Texas Instruments, ITT, Rockwell International, Arthur D. Little, and Gulf Research and Development (Gulf Oil). In fact, 73 of 139 top ERDA executives were "borrowed" from corporations in the energy field, and 55 of those (40 percent of ERDA's senior staff) were from corporations holding current prime contacts with ERDA.

A typical and important example of such interlocks, in addition to Seamans and Fri, is an electrical engineer named David Israel, who in 1977 was appointed assistant ERDA administrator for field operations. Before joining ERDA in 1975, Israel had worked for the MITRE Corporation, the Defense Department, and the Federal Aviation Administration. The MITRE Corporation is one of seven or eight national "non-profit research institutes" whose primary occupation is government funded research on behalf of the U.S. corporate class, particularly the military-industrial complex. (Others include the RAND Corporation, Battelle Memorial Institute, Stanford Research Institute, the Bechtel Corporation, and Midwest Research Institute, which now operates the government's Solar Energy Research Institute in Golden, Colorado.)

Among MITRE's clients are 18 different "commands" in the Defense Department, plus the Departments of Transportation, Commerce, HEW, State, and Justice, the CIA, FEA, NASA, and ERDA. Its clients in the "private sector" include the American Bankers Association, Cal Tech, the Ford Foundation, and EPRI, as well as several electric and gas utilities. The chairman of MITRE's Board of Trustees is Robert Charpie, president of the Cabot Corporation of Boston, with other trustees representing Tyco Laboratories (which spawned, in collaboration with Harvard, a photovoltaics company now owned by Mobil Oil), the First Boston Corporation, Polaroid, Eastman Kodak, Harvard, MIT, Dartmouth, and the University of Michigan.

The MITRE Corporation has received millions of dollars in ERDA/DOE grants since 1975, including over $500,000 for a "fuels from biomass" project in 1976 and nearly $200,000 from the Solar Division in 1977. Employees of MITRE have become so thick in the Geothermal Division that a DOE official named Don Elmer complained in an interview of not being able to reach the Xerox machine "for all the MITRE people standing in line." That same official provided a clue to the origin and function of "think-tanks" like MITRE: "You don't want a university doing (government) research if you can have a 'lab' doing it," said Elmer, "so what you do is the university (or, in MITRE's case, the Air Force) creates a 'lab' and gradually over time it grows until lo and behold it stands on its own feet." He further cited the Applied Physics Lab of Johns Hopkins University, "which was created by the Navy to do research on submarines. Now they get bored and they want more money so they come hustle us." Elmer has nothing but contempt for MITRE and its corporate brethren on the dole from DOE. He calls them the "Beltway Bandits," referring to the beltway around Washington, D.C. which is dotted with their headquarters buildings. "These guys produce bullshit," he said: "Every one of the big boys turns out useless crap...and we keep giving them money." Part of the reason for that, of course, is the presence of executives from these corporations in the signal posts of DOE and other federal agencies. (The man whom David Israel replaced as assistant ERDA administrator, incidently, was Michael Yarymovych, who subsequently went to work for Rockwell International—$1.3 million from two branches of the Solar Division in 1976 and $1.2 million in 1977.)

While the Solar Division was thus not immune to such abuses, the story there is perhaps more wrenching because of what that division, in particular, *might* have done for the country. Until its absorption by DOE in October, 1977, the Solar Division was part of a larger ERDA unit presided over by the

Assistant Administrator for Solar, Geothermal, and Advanced Energy. That slot was originally filled by a scientist named John Teem, who quit in disgust early in 1976 when President Ford's budget director shrank the solar budget for 1977 from $255 million to $116 million. Teem was replaced by Robert Hirsch, a former employee of ITT who had spent the last few years as Director of Controlled Thermonuclear Research at the Atomic Energy Commission. Hirsch aroused the ire of solar advocates in late 1976 by trying to reduce the scope of the projected Solar Energy Research Institute from a major national laboratory with a budget of $50 million by 1980—as proposed by the National Academy of Sciences—to what the *Science and Government Report* called "little more than a management and research team at an existing institution." (The SERI which emerged is somewhere in between.) With Carter's accession in January, 1977, though Schlesinger had guaranteed Hirsch a senior office in the DOE, he resigned to take a job as Deputy Manager of Science and Technology for Exxon. He told the *Energy Daily* that it promised to be a "terribly interesting" position.

The origin of many of the branch chiefs and other key officials within the Solar Division has already been discussed. People like I.O. Herwig (Scientific Advisor), Richard Bleiden (Assistant Director for Solar Electric Applications), and R.F. Morse (Bioconversion Branch) all came from NASA, NSF, and AEC. But the director of the Solar Division, Henry H. Marvin, came from a long career at General Electric, where he managed the High Intensity Quartz Lamp Department, while his Deputy Director, Howard Coleman, had been for the last eight years a private "energy consultant" with numerous clients in the higher reaches of the corporate energy fold. (Coleman turned up in August, 1977 as an ERDA representative to the International Conference on U.S. Options for Long-Term Energy Supply in Denver—sponsored by the Atomic Industrial Forum.) "If solar energy is used widely," Coleman told the conference, "it will have to be done through utilities, and that's where the government is concentrating its demonstration funds." He added that "solar energy will be used as a complementary power source to other generating plants, particularly nuclear power plants." Coleman is now in charge of DOE's solar satellite program.

According to a Texas State energy official who visited ERDA in the fall of 1977, the entire Solar Division "seemed slanted toward the big corporations." Division employees, he said in an interview for this study, feel that "if we're going to have mass production of solar commodities at reasonable prices, we've got to turn the work over to the big corporations. And Henry Marvin is the root of that thinking. All the philosophies of Division staffers, their very phrases, come from Marvin." They contend, among other things, that any small firm desiring government support in the solar field "had better tie in with a major corporation." One of the Solar Division's first substantial contracts in 1975 went to General Electric—a $900,000 grant "to define approaches for carrying out the National Plan for Solar Heating and Cooling" (invoking a certain irony, since General Electric had determined just a few months earlier, at a cost of $500,000, that solar energy by the year 2000 would account for less than 2 percent of the nation's energy supply). In a classic gesture of government efficiency, the Solar Division matched the G.E. grant in 1975 with a $900,000 award to Intertechnology Corporation for exactly the same study.

This combination of corporate control at the top—from the Business Roundtable through the Labor-Management Group to the advisory councils of ERDA and DOE—plus faithful employees in the middle ranks of government has produced what executives call a "favorable costs climate for business." Indeed, according to an article in *Business Week* (June 27, 1977), "General Electric now has over one-third of its central laboratory force of 1950 people working on such diverse energy-related technologies as coal gasification and electric cars, and it figures that energy R&D is growing one-third faster than any other research category in the company." *Business Week* says that "corporate competition for government funding is keen," because "ERDA contracts for R&D are profitable: TRW, Inc., an auto parts and electronics company that is mounting an aggressive bid to enter the energy business, earned around 3 percent after taxes on its $16 million in ERDA research revenues last year. Such contracts entail little risk, and 'there's no investment to speak of—just bright people,' says Richard D. DeLauer, TRW's executive vice-president for systems and energy." But, adds *Business Week*, "the big attraction of government funding is that it offers recipients the chance to build a major new business at taxpayer expense. Such windfalls are common in the defense business, so it is not surprising that defense contractors are elbowing their way into ERDA's receiving line."

TRW commenced its elbowing in 1972 when company managers, according to *Business Week*, "began mulling ways to capitalize on the energy tightness they foresaw." They hired John S. Foster, Director of Research and Engineering for the Defense Department, to come play on their team, and he helped conceive a corporate strategy whose "first major thrust was 'to ship 100 people to Washington' in 1974 as a task force to advise the government on organizing the new ERDA." (This was also the year that TRW submitted *its* $500,000 prediction of a 3.1 percent solar "penetration" of the energy market by 2000.) "TRW's task force," says *Business Week,* "still in Washington, gets paid for such advice, but its main goal is to build relationships with government energy research agencies and sound out bidding opportunities." The TRW strategy "is paying off, at least in outside research contracts. The company will spend some $45 million for energy R&D this year, 60 percent of it government funded (and much of the remainder covered by independent organizations such as the Electric Power Research Institute). The total should grow to $100 million by 1980, DeLauer predicts." He also told *Business Week* that if "ERDA-funded projects go commercial, TRW must simply reimburse the government for any tooling or federal facilities it has used for research—a negligible sum...'There's no question,'" he said, "'there's going to be a major profit opportunity for TRW in energy.'"

How true and how costly to the nation and its small entrepreneurs like J. Hilbert Anderson, who with his son heads a firm in York, Pennsylvania called Sea Solar Power (SSP). Anderson has the unattenuated misfortune to be a competitor with TRW and other corporate behemoths which have "elbowed" their way into the business of government funded research on a solar-electric concept known as ocean thermal energy conversion (OTEC). It is a simple concept, involving the use of temperature differences between the surface of the ocean and its depths (3000 feet) to activate a "working fluid" such as ammonia which in turn drives an electrical generator. The entire apparatus would float on a seaborne platform with a pipe extending into the colder layers of the ocean. This pipe would recycle the working fluid through a cold-water condenser and return it to the platform to pass once more through the generating turbines—rather on the order of a "perpetual motion" machine. Proponents claim that such devices could generate enough power from the Gulf of Mexico alone to equal 100 times the present consumption of electricity in the U.S. with little or no effect on the marine environment. A scientist told a congressional committee in 1974 that "solar-sea-power plants have not attracted the attention of physicists probably because their operation does not require sophisticated physics—only sophisticated plumbing."

Anderson has been at work on his version of the OTEC concept since 1965. He has applied repeatedly for government support, finally squeezing $31,000 from NSF and two grants from ERDA for $39,000 and $144,000—the latter to design a "compact heat-exchanger" that could theoretically be matched with a plant sold by Lockheed or TRW. Meanwhile, Anderson has invested $180,000 of his own money to improve his design and build the only working prototype of an OTEC plant in existence. Mounted on a drum of water whose lower depths are cooled with ice, it generates 150 watts of electricity. The government technical review of Anderson's concept was performed by TRW itself (which evidently has "review authority" over all applications for OTEC funds from the federal solar program—just as the Battelle Memorial Institute has "review authority" over biomass applications). That review concluded that the SSP design would not only work but be cheaper to build than either the TRW or the Lockheed design. An SSP plant of 100MW would cost $150 million or $1,500 per kilowatt (KW) of capacity, while the TRW plant would cost $2,100 per KW and the Lockheed plant $2,600 per KW.

Another study by the Philadelphia Electric Utility compared the SSP design with other sources of electric power in terms of the cost of that power delivered to consumers. The utility found that electricity generated from petroleum in 1980 would cost 44.5 mils per kilowatt hour (KW/hr), 27.3 mils from nuclear, 37.5 mils from the OTEC plant proposed by TRW, and 14.7 mils from Anderson's OTEC plant. A scientist writing for *Nature* magazine has judged the SSP concept superior to the others he has seen because of its "skeletonized design, making its weight about one-tenth that of the TRW design, with corresponding cost savings."

Still, at a September 1976 "energy briefing" at the University of Texas, Anderson decried the fact that "the government and ERDA have largely ignored us." He added in a telephone interview in early 1979 that although he had been awarded a pair of small ERDA grants, plus a subcontract from General Electric to study components for someone else's OTEC plant, "it's just that: a study. We're doing studies of other people's studies which will then be studied by other people and probably forgotten. We still are getting the ugly end of the stick. We still have no money from the government to put in our plant, and that's what counts." Neither has Anderson been able to attract conventional financing because investors are waiting for the government to advance the OTEC concept to a stage that would be "commercially viable." So Anderson cools his heels while TRW and Lockheed, DOE's favorites in the development of OTEC, continue with their corporate buddies to sponge up the millions. In 1974 (the year John Foster's "task force" arrived in Washington), TRW received $391,000 from NSF to design an OTEC plant, while Lockheed got $330,000 for the same purpose. The designs they produced were strikingly similar in basic hardware not only to each other but to Anderson's design.

Then, in 1976, TRW copped $529,000 from the Solar Division for a nine-month study "to develop alternative, non site-specific OTEC facilities and ocean platform requirements for an integrated OTEC test program that may include land and ocean test facilities." In keeping with apparent tradition, Lockheed got $550,000 for precisely the same study, plus $250,000 for an

"OTEC tube and shell heat exchanger producibility study," plus $2,495 to assess the "potential for accelerating commercialization of OTEC." This round of funding also provided $1.3 million to Gilbert Associates, a Washington, D.C. architectural firm, for "architect engineering services in support of the OTEC program." Other recipients of Solar Division OTEC funds in 1976 were: General Electric ($150,000), Westinghouse ($378,000), Union Carbide ($519,000), Institute for Gas Technology ($300,000), Aerospace Corporation ($210,000), Battelle Memorial Institute ($1.5 million), Carnegie-Mellon University ($1 million plus $800,000 in subcontracts from Battelle), Allied Chemical ($47,250), University of Delaware ($52,000), Alcoa ($34,975), Arthur D. Little ($50,200).

But the best was yet to come, at least for TRW, Lockheed, and Westinghouse. In 1978, to supplement the combined total of $539 million awarded to the three corporations by DOE in 1977 alone (primarily for nuclear R&D), the agency granted *each* of the giants a contract in excess of $2.5 million "to prepare preliminary designs for two OTEC facilities—a commercial-scale facility and a scaled-down pilot plant" (*Energy Daily,* May 2, 1978). These contracts repeat the pattern of others cited above in terms of gross duplication of effort. ERDA's huge grant to Battelle in 1976, for example, was to fund a comprehensive OTEC "bifouling and corrosion study." That was also the object of the grants to Allied Chemical, the University of Delaware, and Alcoa (not to mention Battelle's subcontracts to Carnegie-Mellon).

It is difficult to believe that so many experts at such expense are required to solve a problem which, in any case, has surely been solved time and again by naval technicians, bridge builders, metallurgists and other professionals whose jobs routinely confront them with "bifouling and corrosion" of metal in salt water. It is thus difficult not to interpret most of these contracts as very expensive busy-work, paperbound excuses for stalling solar energy while funneling millions of government dollars to aerospace corporations, corporate think-tanks, and corporatized universities. This is thrice true considering the fact that J.H. Anderson and son have a perfectly workable, low-cost OTEC design which they were refining before most Lockheed executives knew anything about solar energy that they hadn't experienced on the sands of the Bahamas. What the Andersons *don't* have are a hundred executives stationed in Washington to "advise" the government on its energy programs and to staff the appropriate government funding agencies.

V

SENATE HEARINGS IN 1975:
SMALL IS BEAUTIFUL

America needs a continual flow of creative entrepreneurs ready to challenge new horizons, but in recent years the obstacles often have been insurmountable. Starved of capital, deprived of incentives, submerged in bureaucratic red tape and surrounded by the burgeoning bigness of the corporate giants, the small business sector has become a victim of the upheavals and recessions of the 1970's. It is matter of national urgency that the small entrepreneur, while not yet extinct, has become an endangered species.

> —Arthur Burck, in testimony before the
> House Small Business Subcommittee on
> Antitrust, Consumers, and Employment,
> November, 1978

The OPEC boycott imposed by the Arabs in October, 1973 caught most Americans largely by surprise. In the winter months that followed—bringing fuel oil shortages, factory shut-downs, and exasperating lines at gas stations—surprise turned to shock and then to a somber realization, at least for some Americans, that the Age of Petroleum was dying as quickly as it had arisen just three generations before. This awareness was buttressed to the point of nausea by a grand display of hand-wringing and pleas for conservation by every politican and concerned celebrity who could wangle time on television. Editorials abounded on the subject. Teachers lectured students and bosses their employees on the virtues of turning down thermostats, donning sweaters indoors, and driving as though with an egg beneath the accelerator. The launching of "Project Independence" in Washington, moreover appeared to enlist the direct support of the people—not just government and business leaders—in freeing us from the tyranny of the oil shieks.

Thus, at about the time that Congress established the Energy Research and Development Administration in January, 1975, there occurred a veritable explosion of hope and activity in the laboratories, garages, living rooms, and offices of small-scale America. Lone inventors and renegade technofreaks, some of them in exile from major universities, unsheathed their calculators and started reading up on solar thermal dynamics and wind behavior. Hippie farmers wrote magazine articles on backyard methane gas digestion. Radical economists and political activists held seminars on a solar-based decentralization of the United States. Small brave businesses were formed—over 3000 by June, 1976—to manufacture and market a diverse array of solar devices, from flat-plate collectors to kits for building windmills and solar greenhouses. By May of 1975, the Solar Divison of ERDA had received more than 5000 inquiries and the National Bureau of Standards more than 600 letters requesting evaluations of solar inventions (7000 by 1979). A year later the Solar Division was still logging up to 200 unsolicited grant proposals per month, while an ERDA-funded Solar Information Center in Rockville, Maryland was processing a monthly barrage of 3500 letters and phonecalls.

Such a response left no question that small-scale America was ready to move on solar power. But the federal government was not at all ready for small-scale America. Sometimes politely and sometimes brusquely, government officials denied the overwhelming majority of requests for support by small solar firms and inventors. Grant applications were returned scarcely opened, or forwarded to corporations and universities for a "technical review" which simply provided access by those corporations to fresh ideas for their own potential use. Other applications were returned with a note that the project in question was already underway by another party or did not fit the needs of the national energy program. Occasionally an applicant was visited briefly by a government official or corporate representative who would nod his head and go back to Washington, never to be heard from again. Still other inquiries brought no response at all. Letters went unanswered and long-distance telephone messages unreturned.

It was this situation, in part, which prompted U.S. Senators James Abourezk, Gaylord Nelson, and William Hathaway to conduct their hearings in the spring and fall of 1975 on behalf of the Senate Select Committee on Small Business. Among the witnesses at those hearings was a Denver businessman and solar inventor named Jerry Plunkett. With a Ph.D. in metallurgy from MIT, Dr. Plunkett heads a firm called Materials Consultants, Inc. (MCI), which specializes in the application of new materials and processes to specific industrial and consumer needs. He is also a director of the American Association of Small Research Companies. In his opening statement to the committee, Dr. Plunkett exclaimed his "alarm that the Federal R&D Establishment has lost its ability to recognize and support innovation of important, but small-scale projects; a sense of frustration that the Federal procurement control cords...developed to protect the public purse have become ropes that bind and, in fact, used to hang small business; a kind of outrage that in the areas of solar heating and cooling, where the Congress specifically directs the special participation of small business, that this directive was all but ignored." (Dr. Plunkett was here referring to a clause in the 1974 Energy Reorganization Act which stipulates that "small business concerns be given a reasonable opportunity to participate...fairly and equitably in grants, contracts, purchases, and other Federal activities relating to research, development, and demonstration of sources of energy efficiency, and utilization and conservation of energy.")

Dr. Plunkett recounted a number of personal experiences responsible for his "outrage." One pertained to a small, highly efficient coal furnace which his company had developed as a clean-burning alternative, installed singly in homes and businesses, to the belching smokestacks of centralized utilities. The invention had won the praise of a former executive engineer for General Motors, who wrote to the committee that in "my opinion, the nation's energy problems can only be solved within the next twenty years by vastly increasing the use of coal. The coal must be utilized as coal—not gasified or the job will never be done in time...Dr. Plunkett's furnace deserves further development." But, said Plunkett, "I have visited a large number of (federal) agencies and yet, to date, have not been able to interest a single person in our technology. Some wanted...a proposal that required a sizeable MCI contribution; another, although eager for a proposal, made it clear that MCI would be asked to forfeit all patent rights."

Plunkett then told of an idea for a "non-solar project to study ceramic materials related to MHD air heater requirements" which he had submitted for funding to the National Science Foundation. He was "informed that this project was of no interest to NSF. Still believing in the merits of the idea, I decided to perform an experiment. I took the plan to the University of Utah. I gave it to them (along with the requisite supporting materials)," and "the University submitted a proposal which was accepted and a project initiated and thus had small business ideas quickly become suitable University research." Since this "project was funded by the same group (a division of NSF called Research Applied to National Needs)" to which he had submitted the original idea, Plunkett demanded an explanation: "In total I have written five letters to NSF-RANN, some of these following visits, but to date their record is unblemished. I have never received a single acknowledgement or response." (Plunkett later wrote the Senate committee that "the University of Utah project was just a little private experiment to confirm my opinions of government prejudice toward scientific research not conducted by large universities or corporations. The University did not know the idea had previously been rejected, and I will say that they have done a good job with the project...I have no ill will toward them; after all, if you give up one of your idea children for adoption you want it in good hands, and I am confident that this is the case.")

Plunkett further testified that he had gone to officials at NSF-RANN to obtain their support for a unique solar "drum-wall" concept which had been developed by a young inventor in Albuquerque named Steve Baer. The device consisted mainly of a south-facing wall on Baer's house assembled from large recycled oil-drums filled with water. "To my surprise," said Plunkett, the concept "was belittled because exposed oil drums were used in the initial application. I was told this didn't look good," and "the system probably didn't work." So Plunkett visited Baer's house, found the system quite effective, and proposed to NSF that at least they fund an instrumentation of the "drumwall" to measure its efficiency. He was told that such a test would have to be conducted by "someone other than Mr. Baer." Plunkett offered to direct the instrumentation himself "or let someone else do it, because it appeared useful to get the data on a new system. Nothing was ever done," and though Steve Baer has identified several important problems and developed various technical concepts, he has not received a single nickel from NSF."

Plunkett had essentially the same experience in connection with a "passive" solar house in Denver. (A "passive" system is one which uses the design features of the house itself, rather than a separate collector and storage tank, to absorb, store, and distribute solar energy.) When Plunkett submitted the idea for evaluation by NSF-RANN, the "response was that 'passive' solar collection was simply a kind of non-technology that was not of interest and probably did not work as well as stated. I did agree that the only performance measurement was the drastic reduction in (the homeowner's) fuel bills and that perhaps a hundred thermocouples would be more convincing. However, when approached to instrument the home, NSF replied that they could not provide instrumentation for every idea that comes down the pike."

Evidently a persistent fellow, Plunkett later sought government support for a low-cost collector system invented by his own company. "Surprisingly, the collector idea was deemed a good one," but Plunkett was told the actual funding would have to come from a different group within RANN. "I made about six phone calls over the following sixty days," none of which was returned, and "two months ago I started calling again, as the party I was seeking had moved to ERDA." A dozen further attempts to reach this party were likewise in vain, so Plunkett gave up. "I often wonder," he said, "if the president of General Electric called if there would be a return call. I have spent thousands of dollars on travel and toll charges, yet I have not even received agreement that I should submit a proposal."

Plunkett's crude treatment was equaled if not exceeded in the spring of 1977 by Steve Kenin, a New Mexico inventor who owns a small firm in Taos called the Solar Room Company. Kenin produces an innovative "solar greenhouse" which the firm both installs and sells in kits for owner-installation. It is a simple structure of translucent polyurethane stretched over a flexible aluminum frame and kept inflated by a small electric fan (16 to 35 watts) which circulates the sun-warmed air of the "solar room" through the rest of the owner's dwelling. Several dozen of these greenhouses, attached to the southern exposures of homes, have long been in use in the mountains of New Mexico. They have withstood 80-mile-per-hour winds and snowstorms while cutting owners' annual fuel bills as much as 50 percent. The "solar room" has been tested and endorsed by Dr. B. T. Rogers, a solar consultant to Los Alamos National Laboratory. It has proven especially popular with poor families because it can be installed for less than $600, thus paying for itself in two to three years.

And yet, when Kenin applied for HUD-ERDA funds to demonstrate his greenhouse on six HUD-administered homes in the area, he was rejected because, according to a technical review, "this is a fragile, ugly appurtenance stuck onto the home, aesthetically bad." Another reviewer complained of the polyurethane, saying it was not durable enough. Kenin—who learned to appreciate simple technologies while helping build low-cost houses in Cuba after Castro's revolution—was incensed. He feels in retrospect that either the reviewers did not study the figures in his application, which Dr. Rogers helped prepare, or were simply not disposed toward something so technically uncomplicated and inexpensive. Kenin admits that the polyurethane "skin" has to be replaced every three or four years, but, as clearly pointed out in his application, the owner saves eight to ten times the $50 replacement cost during the life of the skin. "The reviewers," he said in an interview, "didn't evaluate the system in terms of its gestalt, the people and lifestyles it was designed for." (A prominent New Mexico newspaper printed an angry story on the episode which brought a letter of concern from ERDA, carrying with it the possibility that Kenin will be funded after all—but not through HUD.)

By way of contrast, about the time that Kenin submitted his application to HUD—for $25,000 to equip six homes and monitor them for 12 months—that agency announced the winners of its "Second Round" of ERDA-financed "solar demonstration" grants to underwrite residential installations around the country. One of the largest grants ($216,500) went to Perl-Mack Enterprises, a Denver real-estate development firm that wished to install solar heating and hot water systems in 22 new homes it was planning. The systems it bought were supplied by Solaron, George Löf's company referred to previously, and the cost to HUD was $9,840 per unit. (The homes, at $50,000 each, were sold virtually overnight.) Such relative extravagance had been prefigured at the 1975 Senate hearings when ERDA's John Teem informed the committee that ERDA was considering a solar demonstration program of 1000 units on a budget of $18.5 million, or $18,500 for each solar heating system. This brought an instant challenge from a small California solar contractor named Jim Piper, who had earlier described a project in Palo Alto where he was retrofitting a 220-unit housing complex for $44,000 or $2000 per unit. Said Piper to Teem: "I will make this offer right now. I will cut your $18,500 in half and take your full 1,000 units if you are ready to sign a contract right now." Dr. Teem was not ready.

In his testimony the day before, Piper had expressed dis-
belief at an ERDA plan to spend $109 million of the taxpayers'
money to fund the installation of solar demonstration systems in
300 homes and 50 commercial establishments (averaging $31,114
per solar unit). "You can build a small city for $109 million," he
said. He then pointed out that his company (Piper Hydro of
Anaheim, California) had recently furnished a solar heating
system for the home of a government employee, located just 35
miles from the site of the committee hearings, which had cost the
owner "no more than a conventional (heating) system. *With*
solar, it cost him no more than a conventional system. He had no
add-on, no penalty, no first-cost penalty." Piper further alluded
to a solar installation in a Seattle home which had imposed a
"first-cost penalty" of only $1200. Jerry Plunkett's testimony also
blasted the $109-million solar demonstration budget. He assured
the committee that he "would be happy" to accept such a contract
"at half the price, and I will guarantee those units for 50 years,
and I can promise you that I will be an extremely rich man at the
end of the projects, and they will be valuable projects. I simply do
not think that these programs are cost effective, or can be cost
effective with these kinds of numbers."

Plunkett, Piper and other witnesses before the committee
hacked away at the myth that large institutions, whether
corporate or academic, could or would even try to be more
efficient than small entrepreneurs and "innovators" in the
expenditure of government funds for solar energy development.
"The federal government," said Plunkett, "has what I believe is
an almost incurable habit of undertaking large-scale projects.
Given two equally valid technical responses to a national
problem...the technology that is larger in scale will invariably be
preferred to the smaller more decentralized technology. For
example, as solar energy development started again (following a
30-year hiatus), one of the first concepts that received attention
and evaluation was the 'satellite in the sky' concept as compared
to home heating."

The "problem" in solar energy, said Plunkett, is not a "basic
research" problem: "We don't have to re-invent the wheel. We
don't have to have college professors tell us what the intensity of

the sun is or that solar energy is a workable system...There are workable systems on people's homes and no need for gathering research data." The "problem" in solar energy is one of moving a simple, long-proven technology out of the laboratory into the "marketplace," and that is best accomplished by small private enterprise. "It requires," said Plunkett, "innovation and innovators, not research." Yet "we have found Federal employees unable to understand the solar state-of-the-technology, unable to formulate reasonable plans for moving solar technology ahead, and in fact, engaging in projects designed to keep university professors employed and off the street, and to use study contracts granted to large firms to make solar energy appear long term, remote, and unlikely to respond to our present energy crisis."

Senator Hathaway asked Plunkett: "What is there about this small innovator that makes him so important vis-a-vis the large corporations or the large universities?" First, said Plunkett, citing his experience as an adviser to many companies and government agencies, the innovator "almost invariably comes at the problem from a different point of view." Whereas the researcher in a corporate or government institution will "structure his programs to serve the needs of the corporation or the government agency"—which needs may extend "from production facilities to the kinds of products they have to introduce" in the marketplace—"the individual innovator does not start at that point at all. He starts at some larger point, such as a societal need, or a perceived need...and he has no constraints. He simply starts thinking about the problem. And from that he develops a solution." The innovator represents a "synthesis" between academic theory and practical application of that theory. "Most bureaucrats," said Plunkett, "who deal with small business inventors are technical people," scientists and engineers who "are unable either to innovate or to appreciate innovation as a process or the 'funny people' who engage in the process. Almost none of the U.S. institutions of higher learning...understand that innovation is a very different process than analysis. The idea of synthesis so essential to innovation is not taught since it is so much easier for professors to make up sets of problems for analysis than for synthesis training."

Jim Piper echoed the implications of this when he told the committee that "the key to the use of solar energy" is not further study of hardware items like the solar panel. "As Dr. Plunkett testified," said Piper, "anybody can make a good solar panel. A smart 15-year old science student can make a good solar panel in a couple of weeks." What is needed now is a widespread practical knowledge of how to install and maintain workable solar systems. "If you cannot go out in the field and present a system that a mechanical contractor can install, and that he does not fear (add a fear factor too), you are never going to see solar systems installed. One of the reasons why we believe that small business should be helped is that you are not going to find the National Science Foundation or any university" developing such a practical curriculum. "Why not?" asked Senator Hathaway. "Because," said Piper, "they do not have the knowledge to develop it. [If you look for a course in the United States that teaches you how to be a builder in a university, you will not find one.]The only course that teaches you how to be a builder in the United States is the course of being one. The knowledge that you gain from working every day in the field, solving the problems, is the knowledge you need to apply solar energy." Senator Hathaway: "Does MIT's School of Architecture teach people how to build?" Piper: "I have a hunch that half of the people in MIT's School of Architecture do not know a sixteenpenny from a two-by-four].If I took a man out of the MIT School of Architecture and said there are the materials for a house, they are on the piece of property next door, and here is the plan, build it, I have a hunch that what he would build, you would not want to live in."

Piper was asked later why the major corporations, "vis-a-vis the smaller ones," should not be entrusted with the nation's solar energy program. "The larger corporations," said Piper, "have vested interests right now in maintaining the status quo. If Westinghouse or G.E. came up with a usable solar system immediately, it would harm their profitability in other areas... Westinghouse just submitted a proposal to the FEA for $1.75 billion for three atomic generating plants. If there was a dramatic increase in solar energy systems and a lowering of the need for electrical energy in the future, it is possible that they might not be able to sell those three plants for $1.75 billion. And that has got to be in the mind of the man from Westinghouse who decides whether to support solar systems or not. There is just not the profit, nowhere near it, in solar technology and equipment. It is just not there." (Piper had earlier referred to "a market philosophy that is dying slow and hard which I call 'energy waste for profit.' ")

Among the other issues addressed at the hearings were patent rights and the question of a handmaiden's role for small business in the execution of federal solar contracts awarded to large corporations. Jerry Plunkett, who had previously mentioned being asked by NSF-RANN to forfeit the patents to his coal furnace in exchange for a grant, suggested to the committee that the government's policy on patent rights, at best, favored large enterprises (with batteries of patent lawyers) over small ones. Plunkett recommended a "dual" patent policy: "For large industry," he said, "there should be federal ownership of patents. Already we see large firms using federal solar funds for a large portion of their work, and then using their own funds to develop and patent key items. I think, therefore, that all related patents developed by large firms (under government contract) should become public property. For small firms and innovators I think the federal government should give the inventor the rights, but retain a royalty income of 5% until such time as the funds paid for development of that particular patent have been paid back three-fold."

Plunkett appealed in somewhat stronger terms for the right of small business to remain independent of large corporations in competing for government work. He told the committee of a tendency among government contract officials to suggest to small firms that they grab the coattails of larger firms in pursuit of the money and facilities to develop their ideas. "You will go into an agency," said Plunkett, "such as FEA or HUD or NSF, which has the responsibility for promoting technology, research, and development in our society, and they say, 'Oh, that's a great idea, why don't you sell it to company X.' Well, I don't happen to want to sell it to company X, yet...I am willing to give the government part of the rights to this system, maybe even all of them if they will support it and see it through, but I'll be damned if I am going to let some private company take me at this point on an innovation that is very valuable." Jim Piper evinced a similar spirit. In response to a senator's remark that larger companies were perhaps less innovative than small ones because of the absence of competitive pressures on them, Piper said: "That is why I am a small businessman. I love the competition and I love to compete with big business, and I think most small businessmen do. But they like to compete on a fair basis...If (big business) will compete with me on a basis of product for product, I love it, and I can change and innovate much faster than the big business can. It takes a committee of 50 engineers a year to do what we do in a month."

Piper recommended that the federal solar program, in dispensing R&D funds to large companies "for hardware and system development," release those funds as *loans,* not as grants. "My reasoning," said Piper, "is that if the government is not going to ask for the money back, what incentive will a hardware manufacturer have to *really* develop something useable? Besides, I can't picture a large manufacturer being given my tax money to improve the (solar) technology I've scrimped so hard to develop." Jerry Plunkett emphasized the need for a wholesale change in the attitude of federal bureaucrats. "The small R&D businessman," said Plunkett, "is viewed (by government technical officials) as being less skilled, less successful, and less stable than the employees of larger firms. The usual unspoken assumption is 'bigger is better' or 'if you are so good, why aren't you big?' Even worse is the image of the individual inventor. He is viewed as a mad, narrow-minded nut. In fact, I have heard the word inventor used rather frequently in a derogatory manner." Plunkett reminded the committee that "the development and growth of the United States is due, in part, to the genius of our innovators. From the time of Ben Franklin and his stove, Thomas Jefferson and his patents...Eli Whitney's cotton gin, Colt and his guns, McCormick's reaper, and more recently the Xerox process, the ballpoint pen, the Polaroid camera and the laser—inventors have played an important role."

It follows, said Plunkett, that federal officials "must be taught to respect the feelings of any citizen who comes to them with an idea, good or bad." Plunkett also recommended: (1) "that ERDA establish an affirmative action program for small business in conjunction with the Small Business Administration;" (2) "that specific targets be set for small business procurement levels," which "should be increased for 1980, 1985, and 1990 in regular increments consistent with the nature of the technology;" (3) "that each state establish a solar center that would work with local small business in the adoption of solar heating and cooling;" (4) "that a center be established at a selected point in the U.S. to work directly with individual innovators," a place where "any inventor could go and discuss his innovation or idea in confidence and receive a fair, unbiased evaluation;" (5) "that more personnel be hired at ERDA with a small business background and that ERDA expand a small business office and activities." If "the major thrust of these recommendations cannot be carried out," said Plunkett, "I would (suggest) that the federal government stop all support for solar heating and cooling projects at the producer level," because "government participation to date has been a direct and unfair subsidization of large firms and universities—at the expense of small businesses. I believe these activities have tended to discourage competition, foster monopoly, and reduce the rate of adoption of solar energy."

VI

FEDERAL SOLAR R&D: CREATING SOLAR IN THE IMAGE OF NUCLEAR POWER

Three years after the hearings of the Senate Select Committee on Small Business, there appeared a report by the House of Representatives which bore out the findings of the earlier panel in concrete and inescapable terms. Released in November, 1978 by the house Small Business Subcommittee on Antitrust, Consumers, and Employment, the report was entitled *The Future of Small Business in America*. It revealed, among other things, that "for the period between 1953 and 1973...small firms produced about four times as many innovations per research and development dollar as medium sized firms, and about 24 times as many as the largest firms." Between 1969 and 1976, moreover, while *Fortune 500* companies generated "less than 2 percent of the growth in new jobs," the "growth of hiring among small, high-technology firms (was placed at) between 24 and 40 percent, nearly nine times that of employment growth in other sectors of the economy." Indeed, for that same period, "small businesses accounted for...virtually all the new private sector employment in this country."

Faced with statistics such as these, faced with the eloquence of America's Jim Pipers and Jerry Plunketts asking for a chance to survive, the federal bureaucracy in general and the solar bureaus in particular have remained impervious in their devotion to the corporate elite. That devotion loomed astonishingly clear in March and April of 1978, when both the DOE and the Small Business Administration, according to *Solar Engineering* (June, 1978), lobbied energetically against a Small (Solar) Business Assistance Act being considered by Congress. It was the contention of both agencies, said the magazine, "that the SBA had all the authority necessary to assist small business development in solar energy." The bill passed anyway, establishing a fund of $75 million specifically for loans to small solar entrepreneurs. There was, however, a subsequent catch: only $5 million was actually appropriated for the loan program, and that was swept up by eager applicants in less than 30 days.

Thus was vanquished another threat to big business as usual by solar officials at DOE, who since 1975 have poured over one billion tax dollars into the coffers of America's largest and richest corporations. This has proven deleterious not only to competing small firms but to the people and the economy of the United States. "Without large numbers of businesses in competition with each other," says the House antitrust report, "decision-making becomes concentrated, as does economic power. The result is fewer choices for the American public, both in a commercial and political sense." By monopolizing the options in the federal solar energy program, for example, the corporations and utilities have been able to steer the nation's solar future—consistent with their strategy of 1972-73—almost exclusively toward large, centralized solar applications not dissimilar to those associated with nuclear power, nor much less expensive. "One estimate of the cost to consumers where competition is not allowed to flourish," says the House report, "is $175 billion per year, and other estimates place the cost of concentration and monopoly within our economy at more than $100 billion annually."

This corresponds with the thrust of remarks by a still-caustic Jim Piper in an interview for this study in 1977, two years after the Senate hearings. "You either do your own thing in solar energy," said Piper, "or you do the government's thing. If you do the government's thing, they have an unwritten set of rules and you make your proposal based on that set of rules. One of their rules is that it must be very expensive." Since 1974, he said, the government and the large corporations have contrived in relation to solar "a very good PR program, an advertising program. At first their ploy was 'Anybody who's involved with this thing is crazy.' Then they said, 'It's too expensive and you can't buy one.' Then the line was 'It's an advanced technology and the people who are saying it can be done are wrong because look at all the big corporations that are trying to do it and look at all the money we have to put out to do a system.' Their next step will be to say, 'Well, the only way solar can be successful is to make electricity out of it, so we'll have to give the utilities a big system to do that with.'"

Piper says that the average cost of installing a conventional heating system in a new multifamily dwelling is about $1 per square foot of floorspace ($1500 for an average size apartment). That is also approximately what his firm charges for a solar system, collector arrays and all. "I just looked at the latest round of HUD demonstration grants," he said, "and the grants they've made for relatively big systems (300 units) are double and triple the cost of ours. None of our people would be eligible for a grant because our systems don't cost any more than conventional systems. Even in a single-family house you're only talking about $2000 to $3000 more than a conventional system. It costs you more than that to make out the grant application."

Piper was referring to the "Third Cycle" of HUD "solar demonstration" grants, announced in June, 1977. The vast majority of those awards, totaling $6 million for solar installations in multi-family dwellings, went to building contractors who had made prior purchase arrangements with the solar subsidiaries of eight or ten *Fortune 500* corporations. Exxon's Daystar subsidiary, for example, was awarded a total of $1.1 million in contracts for heating and hot water systems, or 18 percent of HUD's entire Third Cycle appropriation. George Löf's Solaron Company took 11 percent of the total at $670,000, while Revere Copper and Brass, Inc., and Sunworks Corporation (both controlled by the ASARCO conglomerate) received a combined total of $1.3 million, or 20 percent of the HUD cycle. These four companies alone, whose representative building contractors had "competed" with more than 700 applicants for Third Cycle funding by HUD, received 49 percent of the total solar budget, and another 38% went to corporations like Grumman, Lennox Industries, and Miromit (an Israeli firm which was purchased in 1978 by American Heliothermal, one of the growing number of solar manufacturers listed on the New York Stock Exchange).

The solar hardware purveyed in these contracts, as suggested by Piper, was anything but a bargain for the U.S. taxpayer. Solaron's heating and hot water systems averaged $11,000 per unit, with some costing up to $15,000. Exxon/ Daystar systems—manufactured in Burlington, Massachusetts and shipped as far west as Santa Monica, California—averaged $9500 per unit with an upper bracket of $12,500. The costs of most of the other large contracts fell roughly in line with these, while Sunearth Solar Products, Inc. of Philadelphia, one of the few small companies included in the HUD awards, managed to install its heating and hot water systems for an average price of $926 per unit. (This compares with an average of $1102 per unit which Piper charged recently to install solar amenities in a 254-unit apartment complex in Ventura, California.)

ERDA itself, in a similar display of largesse with government funds, announced in March, 1977 that its "Second Round" of commercial demonstration awards (as opposed to residential) included a gift to the Xerox Corporation of $540,558 to retrofit a Xerox office building in Stamford, Connecticut with solar equipment. Another award of $496,000 went to a Coca Cola bottling plant in Jackson, Tennessee—while a fire station in Long Beach, California received $450 toward a $13,000 solar system it plans to install. "That's all they asked for," said Henry Marvin, then director of ERDA's Solar Division, when questioned about the grant by reporters. He also defended the grant to Xerox (by far the largest of the 80 grants awarded), saying that the purpose of the grants was not "charity" but "to get some data on solar energy." The exclusion of facilities like a Xerox office building and a Coca Cola bottling plant would presumably render Marvin's data incomplete. Meanwhile, those two firms have doubtless instructed their advertising agencies to begin drawing up a series of ads portraying their solar systems as further proof that the best solution to the energy crisis is the ingenuity, resources and efficiency of America's corporations. (The "demonstration" contracts awarded in HUD's "Cycles 4 and 4-A," as well as DOE's "Commercial Cycle 3," announced in the summer of 1978, were if anything more coldly biased toward the *Fortune 500* than those discussed above. The 1978 awards are summarized in Chapter XII.)

With Revere and Exxon (among others) busy in the solar collector field, General Electric, Boeing, and United Technology (among others) are applying *their* expensive ingenuity to the task of producing electricity from wind-power. In October of 1976, ERDA signed a $7-million contract with G.E. and the Hamilton-Standard Division of United Technology "to design, fabricate, assemble, install, and study a 1.5-megawatt wind turbine electrical generating system." Two months earlier, *The New York Times* carried a gloomy item detailing the failure of a 100-kilowatt windmill constructed by NASA engineers at Sandusky, Ohio. This giant lemon, with a pair of 62-foot aluminum props weighing one ton each, generated electricity for only 30 hours before being scrubbed at a cost of more than $1 million. Yet, according to the *Times*, the projected 1.5-megawatt machine (15 times the capacity of the NASA device) will "in terms of basic design follow the Plum Brook model" at Sandusky. The newer version will also be mounted adjacent to a utility company where, if it works, it will feed its power into the company lines and thence to the consumer ever hungry for more juice.

This approach to wind power, while perfectly consistent with the corporate-government preference for centralized energy systems, has been seriously questioned by scientists even in the government itself. A study in 1977 by the congressional Office of Technology Assessment scolded ERDA for neglecting the potential of "on-site" (decentralized) power generation. The report concluded that "devices having an output as small as a few kilowatts can be made as efficient as larger devices." And a mid-1977 article in *Science* magazine observed that ERDA had not looked "carefully enough at the prospect for improved versions of small wind turbines for distributed applications, or at the potential economies of mass production that might apply to small devices but not to large ones." The same article referred to a study by a noted British astronomer which found "that a distributed network of small wind turbines provides the best match of potential supply to demand (in England) and would be competitive with coal-fired or nuclear generating stations."

Not only has ERDA (DOE) not supported the development of small wind systems—which once were as common and almost as cheap as salt on America's farms (2.5 million of them in 1935)—it terminated, in December, 1976, one of the few advanced small wind programs then underway in the United States. Previously funded by NSF, the project was located at Oklahoma State University in Stillwater. Its research objectives included the refinement of a turbine/prop concept known as variable speed constant frequency (VSCF), the chief advantage of which, over other concepts, is the turbine's ability to provide a steady output of power in fluctuating winds. A further research target was a unique corollary system for producing hydrogen (as useful a fuel as natural gas) through electrolysis. The OSU design has been lauded by *Windustries* magazine as "far and away the most extensively researched VSCF concept...successfully built and tested" to date.

The "prop" for the wind machine, resembling an oversized bicycle wheel, begins to turn and generate power (30 to 50 kilowatts) in winds much lighter than those required of the "Plum Brook" model to be emulated by General Electric. Moreover it works, as documented by a cover story in *Popular Science* magazine which brought the designers—Professors Ramachandra Ramakumar and Jack Allison—5000 inquiries about the turbine from around the world. But "the coming of ERDA," says *Windustries*, "has meant the end for the time being of the OSU program. Jack Allison is philosophical about the decision. He feels that the concept of energy farming on a family scale, with wind-generated hydrogen producing a storable energy surplus, has been proven, and given maybe five years time, the feds will reconsider 'and then we'll take it off the shelf, dust it off, and go to work again.' Meanwhile, he has found the United Nations, in particular the UN Environmental Program, to be more interested in community-scale wind systems than the United States." In fact, Allison is planning a UNEF-sponsored wind project for a village of 1850 people in Sri Lanka.

As of January, 1978, DOE was charging ahead with its program of subsidies to large manufacturers of large wind-electric machines. Out of a total of $24.5 million disbursed by the agency for wind technology in fiscal 1977, more than $17.5 million was awarded to such corporations and research institutes as Lockheed, Westinghouse, NASA, Battelle (Pacific Northwest Laboratory), Boeing, Grumman, and General Electric. The latter company alone received $5.8 million toward a monumental two-megawatt wind system with a propeller diameter of 208 feet to be constructed at Boone, North Carolina. Another of the large contractors was Rockwell International, which operates the government's Rocky Flats wind-test facility at Golden, Colorado. Here, ostensibly, the smaller wind devices of two small producers—North Wind Power Company and Windworks— were to have been tested and evaluated in 1978, along with comparable systems by Grumman, Alcoa, Aerospace Systems, Inc., and United Technologies. It is to be hoped that the larger companies involved in this evaluation, all of whom are long-term corporate partners of Rockwell International in a variety of other projects, will not be accorded the prior consideration which often is bestowed by one representative of the corporate elite upon another. (At any rate, the bewildering story of yet another small wind producer ignored by government and harassed by corporations—Dan Schneider of Irving, Texas—is recounted in Chapter XI.)

It was noted earlier in this study that the shapers of America's solar energy program had decided by 1973 to emphasize the use of the sun for generating electricity as opposed to systems for direct heating and cooling. The four different concepts then envisioned and budgeted for solar-electric R&D— OTEC, wind-power, photovoltaics, and the "power-tower"— continue to be the mainstays in the government's program (though their claim on the total solar budget had receded from 75 percent in 1973 to 65 percent in 1977, due largely to public and congressional pressure for more attention to heating and cooling technologies). Of those four approaches to solar-powered electricity, the most heavily funded and apparently valuable to the corporations and utilities involved is the "power-tower" or "central-receiver" concept. The Solar Division of ERDA, for example, whose solar-electric budget in 1977 was $160 million— against $86 million for heating and cooling—allocated $79 million for the power-tower alone, with $19 million for photovoltaics, $17 million for wind, $12 million for OTEC, and $4 million for the "Solar Satellite Power Systems" program. In DOE's "five-year budget" for 1979-1983, furthermore, according to a study in 1978 by the General Accounting Office, the share to be spent on solar-thermal electric technologies will total $950 million, as compared with $382 million to be spent for the "heating and cooling of buildings" during that same period.

Among the questions compelled by such a budget—and one deserving mention before a discussion of the power-tower program in particular (with of course its maltreatment of small producers)—is the relative wisdom of trying to adapt a solar energy approach to electrical power generation in the first place. What is clearly implied in the bias of DOE's program, as in the government's nuclear and coal-fired electricity programs, is that America ought to expand its electrical generating capacity before virtually anything else that might be derived from a new energy technology. Yet this assumption, at least on grounds of technological and economic efficiency, has been riddled by critics like Amory Lovins, the young physicist whose "soft-path" energy vision was outlined earlier. In his celebrated essay in *Foreign Affairs*, Lovins makes a distinction between the "low-grade" energy requirements of industrial society and its "high-grade" energy requirements. Lovins argues that low-grade energy, which is readily available from direct on-site consumption of coal, petroleum, wood, solar power, etc., is all that is needed for 92 percent of America's "end-use" fuel demands, including space heating, industrial processes, and automotive transportation (bearing in mind that low-grade doesn't necessarily mean low-temperature). Of the remaining 8 percent, Lovins finds that half is consumed by "industrial electric motors," and the rest, "a mere 4 percent of delivered energy, represents *all* the lighting, electronics, telecommunications, electrometallurgy, electrochemistry, arc-welding, electric motors in home appliances and in railways, and similar end uses which now *require* electricity."

While we might, therefore, painlessly restrict our consumption of high-grade electrical power to 8 percent of our energy needs, we instead apply it to low-grade functions as well (such as drying clothes), which boosts our consumption to 13 percent, accounting in turn for 29 percent of our annual consumption of fossil fuels. (A barrel of oil dumped into an electrical generating plant renders one-third barrel in equivalent energy at the point of use and wastes the other two-thirds.) "Where we want only to create temperature differences of tens of degrees," says Lovins, we should use sources "whose potential is tens or hundreds of degrees, not (sources) with a flame temperature of thousands or a nuclear temperature of millions—like cutting butter with a chainsaw."

He adds that a rigorous program of "technical fixes"—an approach to energy conservation which carefully matches end-use needs to sources of power—could further reduce electricity consumption to 5 percent of our total energy demand, "whereupon we could probably cover all those needs with present U.S. hydro-electric capacity plus the cogeneration capacity available in the mid-to-late 1980's ('cogeneration' being the use of industrial exhaust heat to produce electricity). Thus an affluent industrial economy could advantageously operate with no central power stations at all!" Lovins adds later, with a smile no doubt, that "civilization in this country, according to some, would be inconceivable if we used only, say, half as much electricity as now. But that is what we did use in 1963, when we were at least half as civilized as we are now." More dramatic still is the fact that such "civilized" nations as Sweden and West

Germany consume, on a per capita basis, between one-half and one-third as much electrical power as the U.S.—and West Germany obtains 29 percent of that from the "industrial cogeneration" process mentioned by Lovins.

The point of Lovin's argument, amplified and documented by other proponents of a "soft" energy path, is simply that we do not need and should not build a single additional "central power station" of any variety whatever. Yet the current national energy program—aiming at a possible increase in our consumption of electricity to 60 percent of our energy needs by the year 2000—is devoted not merely to hundreds of new central electric plants but to the idea of powering some of them with solar energy. Assuming now, in the interest of debate, that we *are* well advised to pursue one form or another of solar-electric technology, the question then turns, as it does with regard to wind-power, on whether that should be a centralized or a decentralized technology, and whether it should be developed by aerospace corporations and utility companies or by smaller business and scientific concerns.

The power-tower concept in favor with the government is a penultimately centralized and very expensive system. Its design entails, in a 10-megawatt version planned for 1981 at Barstow, California, the mounting of a 750-ton boiler and an 880-ton generating turbine on a concrete tower 500 feet above a field of 2200 large mirrors or "heliostats." These mirrors will "track" the sun along its daily arc, concentrating and redirecting its rays toward the "central receiver" or boiler on the tower, there producing steam as hot as 1750 degrees Fahrenheit to drive the generating turbine. (The Barstow plant will be operated by Southern California Edison Company, senior member of a consortium which will share about 20 percent of the $130 million which the unit is expected to cost.) In preparation for this facility, a 5-megawatt experimental power-tower is nearing completion at DOE's Sandia Laboratory in New Mexico, where the central receiver and its 200 heliostats cover 100 acres of land. Each heliostat, anchored in ten tons of concrete, has a surface area of 37 square meters and a price tag of $18,500 to $37,000, depending on which of four corporate contractors manufactured it. The ultimate goal of the power-tower program, to be reached in 1985, is construction of a "commercial-size" 100-megawatt unit on a 1000-foot tower surrounded by 10,000 heliostats covering one square mile of territory.

This approach to solar electricity has of course been subjected to the same sort of criticism as the government's and General Electric's draconian wind machines. In the study cited earlier by the Office of Technology Assessment, the authors discern "no clear indication that large solar electric plants are more efficient or produce less costly energy than small, on-site facilities." *Science* magazine, having identified the power-tower as "the concept which the utilities' research arm (EPRI) sees as the most likely candidate for central electricity generation," further observes that "despite the diffuse nature of solar energy, ERDA's (DOE's) program has emphasized large central stations to produce electricity in some distant future" at the expense of "small solar devices for producing on-site power—an approach one critic describes as 'creating solar technologies in the image of nuclear power.' " *Science* refers to Martin-Marietta, Honeywell, McDonnell Douglas, and Boeing as the major contractors in the program, adding that if "the power tower proceeds apace, their names will become as synonymous with solar electricity as the names Westinghouse, General Electric, Combustion Engineering, and Babcock and Wilcox have become with nuclear power." These aerospace corporations, says the magazine, have promulgated "massive engineering projects" which "seem to have in mind the existing utility industry—rather than individuals or communities—as the ultimate consumer of solar equipment." *Science* concludes that such a casting of solar technology in the nuclear mold is at best premature, because "even for the specific purpose of converting solar energy to electricity via thermal systems, there are many choices, and new inventions are appearing rapidly. It would appear to be far too soon for the solar program to be discarding innovative options and sinking its money into steel and concrete."

* * * * *

One of the "innovative options" which ERDA and DOE have striven to "discard" is a remarkable on-site system developed by a small company in Austin, Texas. The firm is Solar Dynamics, Ltd., and the invention to which it has given frequently painful birth is a modular solar-electric device which can be installed on virtually any scale from a two-kilowatt home-size unit to a 500-kilowatt power plant appropriate for a small municipality. The system combines a low-cost, high-intensity concentrating collector with a heat-pipe which carries super-hot metallic gases past a storage bed of molten salts. These salts absorb the heat from the heat-pipe, storing some and releasing the rest to operate a generating turbine. As the heat from the gaseous metal is transferred to the salts, the metal in the heat-pipe cools to a liquid state, returns by convection to the solar collector, is again super-heated, and rushes once more toward the storage salts.

David Marke, the 33-year-old scientist who developed the concept, views his molten-salt storage feature as perhaps the key component in the Solar Dynamics system. It is, at any rate, one of a number of features which make the system not only cheaper but apparently superior to DOE's power-tower. Marke's system, because of the storage feature, can generate electricity for up to 28 days *without sunlight*, while the power-tower scheduled for Barstow will have only *three hours* of storage capacity. (This will necessitate a conventional "back-up" system fired by coal or nuclear power, a requirement not displeasing to the utility executives who have sanctioned the power-tower.) Marke's plant, furthermore, on the 100-kilowatt scale which he has proposed for a small northeast Texas town, would cost less than $1 million to build, or $10,000 per kilowatt of capacity, as opposed to a minimum of $13,000 per kilowatt for the Barstow plant. The efficiency of Marke's system, in terms of the ratio of solar input to electrical output, would range from 25 to 30 percent, while the power-tower will achieve only 16 percent. Maintenance would be cheaper on Marke's system, too, since his collector (the most fragile part of such a device) will cost only $750, as opposed to the current range of $18,500-$37,000 for the power-tower heliostats. Finally, owing to the fact that an on-site system is free of the need for expensive transmission towers and high-lines (not to mention the absence of fuel costs), Marke's plant will deliver electricity at a price to the consumer of only 2.51 cents per kilowatt hour. This compares with an average of 5 cents per kilowatt hour charged by existing utilities and a projected 4.5 cents for the power-tower.

As with Steve Kenin's "solar room" and J.H. Anderson's OTEC plant, the demonstrable superiority and economy of Marke's system have brought him little but grief in his attempts to pry much-needed development funds from the federal solar program. It wasn't until late 1977, after two expensive years of neglect and occasional abuse by government and corporate executives, that Marke and his partner finally persuaded ERDA to look seriously at their concept—and only then with the potent assistance of a top nuclear scientist at Los Alamos Laboratory. A primary obstacle in the company's earlier appeals to ERDA was the danger that Marke's novel design, were it disclosed in an "unsolicited" proposal for funding, would not be kept confidential by government officials. Marke said in an interview for this study that he had been told "by people very high in ERDA not for any reason to divulge patentable ideas" in a funding proposal. Marke's partner Johnny Carter was like-wise instructed by a Solar Division official that "any proprietary information contained in a proposal would be considered fair game." When Carter attempted to debate this point with the official via long-distance telephone, "the guy," he said, "hung up on me."

Another problem for Solar Dynamics was first in learning of opportunities to submit proposals and then in financing the paperwork required by such proposals. "The way you find out about a Request for Proposals (RFP)," said Marke, "is to read the want-ads in the *Wall Street Journal* and other big dailies. If you're a major company, you can hire someone to just sit down and read the goddamn things. But a company like ours, with two or three people on board, we can't read the goddamn news-papers everyday." While the government is supposed to send out a schedule of RFP's to everyone on its mailing list, said Marke, "we haven't been able to get one of those lists except outdated by a quarter. We're supposed to be on the list to receive them, but we don't get them." Even if they did, they would be hard put to respond because of the cost involved. "The last RFP we turned down," said Marke, "would have cost us $25,000 just to submit the proposal. It would have taken three or four thousand professional man-hours." And that, said Johnny Carter, was among the cheaper RFP's the firm had confronted: "Some RFP's are an inch thick. They want to know who you're going to work with, where they went to school, what kind of computer you're going to use, where you're going to get the computer—it's incredible." Carter added that Solar Dynamics must compete on RFP's with companies like Boeing and McDonnel Douglas. "They can afford to sink $100,000 into paperwork," he said. "We can't."

But the single most intractable barrier to government support for Solar Dynamics, according to both partners, was and is the prior commitment of federal funds not only to the power-tower but to those corporations and universities who already have, as it were, their feet in the concrete. This commitment, said Carter, "has even prevented us from dealing with certain corporations." Among them are Babcock and Wilcox and Rockwell International, who in the spring of 1977 dispatched teams of senior engineers to Austin to evaluate the Solar Dynamics concept, presumably with the idea of a possible joint venture. Both teams praised the concept, then left town and sent word to Solar Dynamics that any potential deal was void because of the larger firm's involvement in the power-tower program. Carter is now suspicious of the companies' original motives: "They deceived us into thinking they were interested in a joint venture so they could come down and investigate our technology." Marke points out that Solar Dynamics is "not competing with that power-tower technology at all. We're set up for modest distributed systems, small-town deals, lower investment." But officials in the Solar Division, he said, "can't see a goddamn thing but the power-tower."

What little else the government will consider by way of solar-thermal electric systems is just as frozen in prior commitments as the power-tower itself. Marke described a 30-minute, between-flights meeting with a Solar Division official at the Dallas-Fort Worth Airport in August of 1977. "This was the *most* direct contact we had ever had with ERDA," said Marke, "and it turned out the guy was working only on systems designed for 200 to 500 kilowatts. If you couldn't meet that capacity, if you had a megawatt or a 100-kilowatt system, you couldn't deal with him. He couldn't *see* anything else. It wouldn't have mattered whether it was us standing there or some guy with a perpetual jelly-bean fountain." Marke and Carter both emphasized that their myriad efforts to communicate with ERDA up to that point had been aimed not at winning funds but at getting some flicker of recognition that they and their system existed. "After the meeting at the airport," said Carter, "we realized that we'd been led on another snipe-hunt. We had spent a lot of money and time and accomplished nothing."

This was the culmination of a three-year struggle for the firm's survival. The partners—starting in Utah with what Carter describes as "a $50-bill and a '65 Ford"—had borrowed money from friends and relatives, gone for months without pay, and somehow managed to stretch the $70,000 they had raised to lease and equip a warehouse with enough sophisticated, second-hand equipment to make good progress toward completion of a working prototype of their system. They were tired. They were sick of the unreturned phone-calls and bureaucratic dead-ends, but they never considered giving up. "I was angry," said Carter, "and more determined than ever that the feds were going to *do* something about Solar Dynamics. We wouldn't roll over and play dead."

It was roughly then that Marke established contact with the nuclear scientist at Los Alamos. He was impressed by Marke's system and began a campaign of personal intervention on its behalf in Washington. So did an official of the Texas Energy Advisory Council. And finally, in December of 1977, an ERDA executive visited the Solar Dynamics facility in Austin and promised the firm a shot at a $100,000 contract to develop its system for home-scale applications. This contract—if and when it comes—may bring Solar Dynamics as many problems as it solves, due to the government's patent policy. The company is not in business for charity. It expects to make a profit from its risks and its 16-hour workdays, and the extent of that profit will be largely determined by the extent to which the firm retains control over the technology it has developed. "ERDA's (DOE's) patent rules," said Marke, recalling Jerry Plunkett's Senate testimony, "are a bit more liveable for big companies (because of their financial ability to secure patents and prosecute violations), but for small companies the policy is murder. If you had obtained, say, four patents prior to ERDA funding, you have to give them license to those patents. Any patents you develop under ERDA contract *they* own and will give you license to, but in no case will you have rights outside this country. So even if they fund you, they've got you so tightly buttoned up that they just stifle the free enterprise system."

Should Solar Dynamics in fact receive the $100,000 for which it has bid, it will perhaps be the only small firm in the United States to have chinked out a niche in the government's solar-thermal electric program. The top grantee in 1977 was Martin Marietta Corporation, whose work on a "central receiver" for the power-tower cost the taxpayer $8.5 million that year. Most of this was to fund a continuation of projects which the company has been pursuing at least since 1976. The same is true of other large corporations receiving power-tower funds for 1977, including Honeywell and McDonnell Douglas, each of whom was awarded a total of $5.7 million. The Aerospace Corporation received $2 million, while Boeing picked up $1.9 million and Rockwell International $140,000. There are other major contractors on the power-tower, including the University of Houston and Sandia National Laboratory (AT&T), but the ones listed above constitute the nucleus of corporate funding recipients. Together, in 1977, they accrued a total of $24 million for R&D on the power-tower alone, which is roughly equal to the government's entire 1977 budget for wind research.

An aide to a U.S. Senator who was interviewed for this study is both cynical and outraged about the massive investment in the power-tower program. "There is some fraud—I mean, there is some stuff in there," he said, "which just drives me nuts. They (ERDA officials) are throwing $5-million chunks around to their friends and neighbors for great arrays of heliostats. Why the hell they need more than a band of five to test the capacity for tracking, reflective power and so forth I'll be goddamned if I know, but they're building them in the range of 36 to 40 units. Aw, Christ," he continued, "ya know what? ERDA's being addressed up here now as the new technological porkbarrel. It's ridiculous the way some of these guys are ripping the country off. There is no contract management over there—it's carte blanche."

In response to a question pertaining to average profit ratios on federal R&D contracts, the Senate aide sighed: "Look. I've spent enough time in the aerospace business, working as a consultant for DOD and the other side—you're *never* gonna be able to penetrate far enough to find out what sort of theft is going on. I mean, it's all loaded into overhead. You've got to go about three layers down in the accounting before you find that all the funds are missible, but, my God, they're making a hell of a lot more money than they were before. It's called overhead, but it's not overhead at all—it's overhead, plus profit, plus bonus, plus you know, all the sweetheart contracts that normally go into that kind of deal. Forget the rate of declared profit. Look at the overhead. I can remember one group getting 135 percent I think it was—overhead on direct labor. They paid their people and they added their bonuses in, and then they hit that by 238, and that was what they billed." The aide sighed again. "People keep saying that we don't have an energy policy, but we have, and we've had one for a long time. I just described it."

VII

By Any Means Necessary: Selling the Corporate Point of View

It will be useful to recall at this point that corporate involvement in solar energy development, despite the appeal of millions of dollars in government contracts, is a second-best proposition in the thinking of most key members of the corporate elite. As mentioned in the treatment of their strategy with respect to solar, most of the U.S. corporate leadership would prefer that such a "radical" energy idea—with its obvious socio-economic implications—had never crept out of the garages and imaginations of America's tinkerers, independent scientists, and small entrepreneurs. In fact, it was and still is a dimension of that strategy to prevent a rush of public demand for solar power, especially of the sort which lends itself to decentralized applications, until the corporations and utilities have (1) completed the expansion of their nuclear/coal-electric grid, (2) negotiated the profit potential remaining in fossil fuels, and (3) hardened their control over the neophyte solar industry itself. This has led to a seemingly paradoxical double-life for those corporations and government agencies obliged to behave on the one hand as proponents of solar energy, while on the other as guardians of a fossil-fired energy status quo. It has led, indeed, to a series of distortions, deceits, and manipulations ranging from the pathetic to the socially dangerous and punishable by law.

Perhaps the least felonious and yet the most effective tactic has been a stream of national advertising and other propaganda, some of it sponsored by the government, proclaiming that cheap and efficient solar energy is at least a quarter-century away. This was Exxon's approach in a series of ads it printed in the fall of 1976 in publications like *The New York Times* and *Newsweek*: "EXXON ANSWERS QUESTIONS ABOUT ONE OF THE NEWEST SOURCES OF ENERGY UNDER THE SUN— THE SUN!" Exxon's "answer" to a "question" about the cost of solar energy was that it "can cost you anywhere from $8000 to $20,000" for home space-heating and, for domestic hot water, "Around $2000—about three times more than conventional systems." As for when solar power will "become a major source of energy," Exxon's tiger says: "Possibly in the next century."

So flagrantly distorted were these ads that they drew a stinging letter of complaint from U.S. Senator Gary Hart and Representative Richard Ottinger to Exxon president H.C. Kauffmann. "We feel," wrote the congressmen, "that Exxon has misinformed the public about the current status of solar energy and the promise it holds for our nation. The answers to questions posed by Exxon in its ad are riddled with innaccurate statistics and pessimistic projections about solar power." The congressmen then rebutted the statements in the advertisement point for point, mentioning that one of Exxon's own solar subsidiaries, Daystar, was marketing hot water systems for half the cost which the oil company had claimed. "In conclusion," they wrote, "we are left to wonder whether the answer to the final question (in the ad), 'Why is Exxon involved in solar energy?' might simply be that Exxon intends for solar energy to be kept under wraps until fossil fuel markets are exhausted."

Kauffmann's reply to the congressmen was unapologetic. In fact, he was able to use the vagaries of government officialdom—theoretically under the legislators' own purview—against them. He simply pointed out that the solar heating systems installed in the first two "cycles" of HUD demonstration grants had indeed cost the government an average of $30 per square foot of collector surface—or $9,000 to $18,000 per housing unit. "This," said Kauffmann, "seems to indicate pretty close agreement with our numbers." He didn't of course refer to his affiliation with the Business Roundtable, nor to the presence of his company's executives on 15 federal energy advisory committees, nor to the employment of former Exxon executives in the very government bureaus responsible for those exorbitant HUD awards in the first place.

The Hart-Ottinger letter also referred to the issue of "life-cycle" cost analysis of solar investments. "Exxon never mentions," wrote the congressmen, "that (because solar units incur no fuel expense) the costs associated with solar energy must be spread over the economic life of the unit." This notion of "life-cycle" costing, as applied in solar energy parlance, stands in opposition to the concept of a "first-cost penalty," and it expresses a crucial distinction indeed—particularly since the latter terminology is brandished so often with precisely the malevolence intended by Exxon. A "first-cost penalty," as the term implies, is that amount which a homeowner must pay above the cost of a conventional heating system fired by gas, oil, or electricity. It is thus the amount that must be "paid back" to the owner in fuel savings in order to justify the solar investment. While Exxon didn't explicitly refer to a "pay-back period" in the ads described above, it is frequently portrayed by other detractors of solar energy, including government officials, as ranging from 10 to 20 prohibitive years.

Thus, in a 1976 propaganda release by Shell Oil Company, the "initial cost" of a solar heating system "for a typical Midwest or Northeast single family residence" would vary "from $4500 to $8000." Assuming that the home in question is presently heated "by relatively expensive electricity," says Shell, it would still take "*at least* 10 years of savings in heating costs to equal the (solar) investment." Furthermore, if the home is presently heated "by much less expensive natural gas or oil, even with escalating prices for these fuels, it will take at least 15 to 20 years before (the solar system) begins to break even." Shell concludes that "the average home purchaser or developer is not likely to view such long payout times favorably." (Shell also concludes that solar energy by 1990 can be expected to supply about "one-half of one-percent of our nation's energy needs.")

Such projections of immense "first-costs" for solar installations serve a two-fold corporate purpose. They help to dampen
popular enthusiasm for solar energy while permitting those
corporations which have entered the "solar market"—like
Exxon—to sock the government and other solar customers with
prices and "payback periods" in fact as unattractive as the
corporations have claimed. Jim Piper, meanwhile, in a 1977
interview with *New Times* magazine, says "Screw payback!" A
"payback period" of "three or four or even seven years is no
measure of whether you should have (a solar system) or not."
Piper says the important economic issue is "return on investment. That is the guts of the argument. Look, my (solar) system
will cost any homeowner from $1500 to $3000 more than a
conventional system. Let's say you buy it for $2000 extra. If
you live somewhere colder than California, your average utility
bills are $40 a month. I'll save you half of that," or $240 per
year. Assuming you have financed your system with a $1600
loan at 8 percent interest, your monthly mortgage payments are
$14. Since you are saving $20 per month in utility bills, you will
pocket $6 per month in "immediate payback"—an 11 percent
return on your investment. "Does your savings and loan do that
well for you?" asks Piper. "And that's with gas prices as they are
now. If the price of gas doubles in the near future, as it well
might, that will mean a 22 percent return on your investment."

The above examples of "solar advertising" by Shell and
Exxon are typical of ads distributed by scores of other U.S.
concerns in the energy field. Still other companies and organizations have slandered solar by failing to mention it at all in
certain contexts where it would be appropriate. The U.S. Chamber of Commerce, in a full-page ad which ran primarily in
business publications in the fall of 1977, touts a six-part "energy-
education" package of slides, cassette tapes, and "evaluation
worksheets." This package, which the Chamber calls a "legis-
lative/education action program," is designed "to demonstrate
the need for positive action by our legislators—local, state, and
national—to improve the energy situation." Entitled "Energy Is

Your Business," the program is advertised as "an overview of the entire range of energy problems," but the only resources in fact discussed are coal, oil, natural gas, and "Nuclear Power: A Passport to the Future." In return for his or her $200, the consumer of this item gets no reference to solar energy. Likewise from Gulf Oil company in October, 1977 comes a packet of energy materials by mail, including a dandy little cartoon booklet on the "energy crisis," which leaves the impression that solar energy simply doesn't exist. The booklet does put in a plug for "strip mining," "offshore drilling," and "a nationwide energy program" to "eliminate price controls," "make more public lands available," and "create incentives" for expanded development of fossil fuel resources—among other suggestions.

Not infrequently, as the following example and others will show, the corporate publicity campaign against solar energy has been funded and tacitly supported by the government itself, in this case the Solar Division of ERDA. Since about the middle of 1974, starting with a seed-grant of half a million dollars from NSF, the Honeywell Corporation has toured the country in a "Transportable Solar Laboratory" (TSL), a large van equipped with a variety of solar hardware, literature, and exhibits on solar energy. The contracted purpose of the TSL and its engineer-driver (for which Honeywell received $1.1 million from ERDA in 1976 and 1977) is to park the van in major U.S. cities, there attracting crowds and conducting workshops on the virtues of solar energy. It has attracted crowds—up to 3000 people a day—but the TSL has not only not promoted solar power, at least not as a currently viable alternative, it has systematically and arrogantly discouraged its acceptance by the public.

During a two-week visit to Sacramento in June, 1977, according to employees of California's Solar Energy Office who wrote an angry memo to their boss, the Honeywell man "emphasized to his audience that solar energy equipment is tremendously expensive. He implied that inordinate amounts of hardware (similar to the variety on the TSL) were necessary for a 'proper' solar system. He used extremely high cost figures" to demonstrate that "applications of solar energy systems (including solar water heating) will be a *net loss* over a twenty-year period." The Honeywell "instructor" remarked at one point that "a smart businessman will continue to buy conventional forms of energy while they are cheap. It makes more sense to save the money that you would invest in a solar system. Put it in the bank and wait ten years." He particularly demeaned "passive solar systems," equating them with "caveman technology...a window on the south side of the house. People who live in passive houses are very different than you and I. Sometimes they are very hot and sometimes they are very cold." This brought laughter from the crowd.

Perhaps the most damning of the "instructor's" remarks was his revelation that "I represent the government more than I do Honeywell." He then provided his audience with a "short lesson on how to win a solar grant from the federal government." One of his tips was to "form a team with a utility. Get them involved." Then: "Get a prestigious architectural firm to work with you. The feds want to see big names working on solar projects." Then: "Get an aerospace company to do your calculations. The feds will be impressed," and "go big. They like to fund big projects."

Lee Johnson, a writer for *Rain* magazine in Eugene, Oregon, says in the October, 1977 issue that the Honeywell abuses in California were brought to the attention of Rhett Turnipseed, an official in the Solar Division, and "the ERDA response over the telephone was appalling." First, says Johnson, Turnipseed issued "a diatribe against the young staffers (of California's Solar Energy Office) who first called attention to the blatant, anti-passive solar, pro-big business/utility bias of the ERDA/Honeywell Transportable Solar Lab." Turnipseed informed Johnson that the staffers' "critique of the TSL could be dismissed out of hand, because (1) they were simply rebellious pro-solar 'young Turks' with 'counterculture connections'; (2) they have no engineering degrees and therefore cannot possibly know enough to critique a certified engineer; (3) Dr. George Löf says their arguments (in defense of passive and low-cost solar energy) are 'technically incorrect'; (4) Löf is advising ERDA on changes to be made."

Johnson replies that the "young Turks" in Sacramento, whom he knows personally, are experienced solar technologists, and one of them, Jeffrey Reiss, "is more knowledgeable on solar than many ERDA bureaucrats I have met." Furthermore, says Johnson, "solar heating methods are too simple and easily understood for the usual technological awe and mystification purveyed by engineers" to have their customary effect. And finally, "it is insufficient and an obvious conflict of interest for Dr. Löf to call Reiss 'technically incorrect' and do the revision of his own TSL workshop material. Löf is an expert, but *only* in expensive, complicated, active (solar) systems such as one finds in the Colorado State University ERDA-funded houses. It would be more appropriate to have an enthusiastic and experienced proponent and builder of passive solar review the workshop materials."

While the ERDA/Honeywell collaboration has obviously damaged small-scale solar energy, a more sinister form of damage was exposed through an incident within the Solar Division itself in the fall of 1976. This misadventure could be termed "The Benson Affair," after Jim Benson, the young employee of the Solar Division who lost his job for having allowed an ERDA contract team to produce a report depicting solar energy as a technically feasible and desirable replacement for nuclear power by the year 2020. The incident reveals quite clearly how far the corporate elite will go to squelch a deviation from its "hard-path" perception of the nation's energy future. Not only was Benson fired from his job, the report he had funded—at a quarter of a million dollars—was first barred from publication and then, due to a spate of unflattering publicity for ERDA and threats of congressional investigations, released with abridgements significantly altering the spirit and intent of the original draft. Among the agents in this intrigue were Henry Marvin, Director of the Solar Division, Marvin's boss Robert Hirsch, now with Exxon, and most or all of the corporate members of ERDA's General Advisory Committee.

The affair started innocently in the spring of 1976, when Benson, who had just been hired to direct the Environmental Assessment Branch of the Solar Division, negotiated a contract with a team of young writers at the Stanford Research Institute (SRI) to project a series of broadly differing "energy scenarios" for the United States through the year 2020. Benson's objective was a comprehensive, statistically reliable comparison of the social, economic, and environmental effects of an "energy scenario" emphasizing continued exploitation of nonrenewable fossil fuels, especially coal and uranium, versus a scenario emphasizing renewable resources in general and solar resources in particular. Much care was taken on behalf of consistency with previous ERDA research. "We used ERDA scenarios," said Benson in an interview for this study, "a very establishment model, and we used ERDA national laboratory environmental data to compute the social and environmental impacts (of each scenario). There wasn't much speculation in there." Some of the data were derived from energy models and projections compiled by the Ford Foundation, Gulf Oil Company, and the MITRE Corporation.

The "reference scenario" in the SRI study was lifted intact from ERDA 49, a document produced in 1975 as a detailed guideline for the agency's solar R&D program. That document is premised on three fundamental assumptions: (1) U.S. energy demand by 2020 will more than double from the current level of 73 quadrillion Btu's (quads) per year to 180 quads; (2) this astonishing leap in demand (which is seen as necessary to maintain the economic growth required by the corporate industrial system) will be met in part by vast increases in the use of coal and nuclear power, the latter of which will rise from a present contribution of 2.2 quads to 30 quads by 2020; (3) renewable resources, with a heavy tilt toward solar-electric power generation, will provide 10 to 12 quads by 2000 and 45 quads by 2020. All three of these assumptions were brought under sharp and plainly worded scrutiny in the original draft of the SRI study, and ERDA's assumptions regarding the environmental and social effects of its scenario were decimated by the Stanford researchers.

"In effect," said Benson in the interview, "we wound up looking at ERDA's entire range of scenarios, including nuclear, coal, geothermal, and so forth. And of course when you do that, the solar impact (on the environment) is negligible compared to the coal and nuclear impact. Solar came out really looking good." It came out looking so good, and the ERDA scenario so bad, that the closing chapter of the doomed SRI report opens with the statement: "*Implementation of* ERDA 49 *will be extremely costly to the society economically, environmentally, and in terms of the necessary implementation steps*" (emphasis in the original). SRI found that ERDA's projections of expanded coal development would bring "severe environmental hazards and the potentiality of conflict between coal producing states and the rest of the nation." The nuclear component in *ERDA-49*, if realized, would be accompanied not only by economic strains and risks of eco-disaster but "by enormous political conflict." And to wait until the year 2000 to commence a tripling of solar facilities by 2020 "would require subsidies in the tens of billions of dollars per year," while the solar-electric dimension of the plan, based on the power-tower, "would involve covering with solar collectors an area somewhere between the size of Delaware and Massachusetts."

The most controversial aspect of the original SRI report appears to have been its strong implication that a U.S. energy-demand of 180 quads per year cannot be met by *any* mix of fuel sources without incurring the likelihood of social and ecological devastation. This leads the authors to suggest heretically that America may well be required to scale down its economic expectations, shift its industrial base from high to more benign technologies consuming fewer finite resources, and—perhaps most insulting of all to corporate eyes—begin replacing its urban metropoles and centralized economic institutions with a new decentralized socioeconomic and political system. The authors refer to the "desirability of a *husbanding* approach toward resources, land, and ecosystems, much as a responsible landed family takes to its estate over many generations." This would imply "an ethic of frugality as regards energy and energy-related goods and services." It would also imply the "progressive re-

design of communities to be more self-sufficient and more con-
genial to the natural environment... and of work-home relationships to require less transportation." Given such adjustments in
the social order, "the advantages of solar energy are so com-
manding that the decision to replace other sources by solar
energy should be made weighting sociopolitical considerations
more than economic ones."

The draft report with these and allied observations, which
Benson submitted to his superiors in August of 1976, hit ERDA
like a bomb. "My boss," said Benson, "called me into his office
and raised it over his head and swung it around and said it was
the most dangerous, un-American report he'd ever seen in his
life, and he wanted me out of there by five o'clock that day."
Benson's boss was Fred Koomanoff, a former executive at
Westinghouse who had recently been hired by another Westinghouse man in the Solar Division named Ray Field. Koomanoff
further told Benson that he "wasn't being a team player and
should be more loyal to the Ford Administration's policy."
Benson now believes, however, that the signal to fire him came
from an office much higher than Koomanoff's—and possibly
from the General Advisory Committee.

The actual firing occurred some two weeks after Benson
had distributed copies of the draft report to Koomanoff, Henry
Marvin, and other officials in the Solar Division. Just a week
before the firing, moreover, Benson had been asked for a Xerox
copy of the report by a secretary to Robert Hirsch, who was
then Assistant ERDA Administrator for Solar, Geothermal,
and Advanced Systems, of which the Solar Division was a
component. Hirsch at that time was engaged in an attempt to
cut back funding for the solar program. "He had worked out a
presentation," said Benson, "that tried to show that solar was
way overrated, and the whole program needed a review to get it
back in balance—in other words to cut its budget. He had gone
to the General Advisory Committee with that presentation"—
having by then seen the Benson/SRI report—"and he also gave
the presentation on Capitol Hill. Here was a guy in charge of
not only solar but also fusion and he thought solar was overrated." Benson added that the General Advisory Committee "is

predominantly nuclear-oriented. It would be impossible to find anyone who is pro-solar on there."

In the weeks following his dismissal, said Benson, before he had left ERDA altogether, "rumor had it that because there was one scenario in the SRI report that showed nuclear being entirely replaced by solar by the year 2020—that was the cause of the uproar. To see the possibility that nuclear could shrink down to less than what it is today by the year 2000 because of solar coming on so rapidly. This was seen (by Hirsch and the GAC) as here's solar competing with nuclear, not only in the scenarios but within ERDA, and we can't have any competition between the technologies, so we better re-evaluate solar." Still later, with the Washington media printing stories about the incident, Benson was summoned to the office of Hirsch's deputy. "He told me," said Benson, "that I should stop talking to anyone about it. If I continued answering questions about the report, my motives would be questioned. I said, 'What do you mean, my motives?' He said, 'You know what I mean.'" Benson's efforts to reach Hirsch himself were futile. "He never responded to my requests to see him personally. I tried to talk to him a number of times, but his secretaries always said he was very busy."

Henry Marvin's role in the "Benson Affair" seems to illustrate the true nature of power-relationships along the corporate-government axis in Washington. In the months prior to completion of the report, said Benson, he had spent many a lunch-hour in Marvin's office chatting about "problems of changing values, depleting resources, the Herman Kahn sort of super-growth orientation versus the Amory Lovins sort of thing. He (Marvin) really, I think, started to understand," said Benson. "We'd go to a staff meeting and he'd repeat things I had said." Once the storm broke over the SRI report, however, Benson was unable to learn whether Marvin had even read the document, much less to elicit his feedback. "Everytime I passed him in the hall for the next two months while all this trouble was going on he would sort of indicate to me that he hadn't read it yet. I just found that unbelievable," said Benson, "and finally came to the conclusion that he *had* read it and was probably under a lot of pressure from Bob Hirsch's office. But he wanted to keep himself above it and clean of it so he let the middleman, my boss Fred Koomanoff, do the dirty work.

"I was taken off the project," said Benson, "my responsibilities were all taken away from me, and the contract expired without the project being finished." Eventually, though, ERDA was forced to finish the project and the final report (entitled *Solar Energy in America's Future*) came out about six months later—down from 400 to 100 pages. The whole scenario had been reworked, rewritten. It was very much toned down." In particular, he said, a discussion in the draft of the need to act *now* in favor of renewable energy resources—before the nation is irrevocably committed to armed nuclear enclaves and gutted Montana coal lands—"was entirely misinterpreted." Perhaps typical of omissions from the published report is this concluding statement in the original: "Solar energy is the *one* energy source for which there are no serious technological gaps, no serious environmental problems, no likelihood of organized public-interest opposition, and no implementation barriers except some economic and institutional factors which can be easily altered by appropriate legislation and procedural changes."

It should be noted, incidentally, that Benson holds no grudges regarding his fate in the Solar Division. He blames his demise and other shabby doings at ERDA primarily on the corporatized Washington atmosphere engendered by the "conservative" administrations of Nixon and Ford. He said in the interview that he expects a change for the better during the tenure of President Carter and his Secretary of Energy James Schlesinger. Yet recent developments suggest the contrary, at least in terms of the Carter administration's willingness to falsify data and suppress information in favor of an energy program supported by the corporate elite. Numerous principles in Carter's National Energy Plan, for example, and particularly those relating to expanded development of fossil and nuclear-electric resources, were allegedly based on a CIA report of April, 1977 which forecasts perilous shortfalls in the world supply of oil, especially from the Middle East, by 1985 at the latest. The report hints darkly at the possibility of having to "compete" for waning petroleum with the Soviet Union and China, since those countries, too, will be drying up. "Before 1985," says the report, "the USSR will probably find itself not only unable to supply oil to Eastern Europe and the West on the present scale but also having to compete for OPEC oil for its own use." In China, meanwhile, "the reserve and production outlook is much less favorable than it appeared a few years ago."

Hence we are warned in Carter's NEP that the "prognosis for the United States and the world is serious if current growth in demand continues. In the short term, American vulnerability to a supply interruption would increase. By the mid-1980's, the U.S. could be vying for scarce oil against its allies and other consuming nations, including the Soviet Union." This means: Don't any of you bleeding-hearts stand in the way of oil conservation measures like nuclear power plants and rising oil and gas prices. But attentive journalists and other interested parties quickly sniffed a skunk in the CIA report, which blatantly contradicted the agency's own earlier estimates of a likely global *surplus* of oil in 1985. Then came an item in the *Wall Street Journal* on April 22 averring that the skunk was Schlesinger himself, who, as former director of the CIA, had requested the agency report "specifically to support the President's new (energy) proposals." (The *Journal's* allegation was later confirmed by CIA Director Stansfield Turner in testimony before a congressional committee.) Nor was this evidently an isolated example of Schlesinger's penchant for tampering with numbers. A December, 1977 study by a congressional task force uncovered 21 instances where Schlesinger had either ordered or approved substantial alterations in data provided by the Office of Energy Information and Analysis (OEIA) for use by the Carter Administration in pursuit of its energy strategy. These revelations were the more disturbing in view of the fact that OEIA had been established in 1976 precisely to foil the statistics-bending long practiced by the oil and gas industry in its provision of critical data to government agencies.

VIII

Utilities and Solar: Toward a Corporate Electric World

Among the most vigorous participants in the corporate effort to control the emergence of solar power, as suggested in an earlier reference to the founding of the Electric Power Research Institute (EPRI), are America's utilities and utility associations. Crisis-ridden since the late 1960's, the utility industry has perhaps more to lose from a popular and "premature" introduction of solar technology than any other sector of the corporate economy. Quite apart from the industry's role in a larger corporate strategy toward a capital-intensive "hard-path" energy future, the nation's utilities need desperately to expand their existing power systems merely to survive their own current financial woes. Utility profits have been seriously threatened by the last eight years of inflation generally, and in particular by a quadrupling in the cost of oil and natural gas consumed as boiler fuels by electrical power plants. These unexpected cost escalations have further strained the industry's ability to meet its obligations on the debt incurred for its massive investment since World War II in plant and transmission facilities. (The utility industry is second only to the petroleum industry in dollar value of capital investments.)

115

Due to its monopoly status and therefore its "regulation" by government agencies, the only way the utility industry can demand the higher rates it now badly needs is to expand its operating facilities, because the rates it may charge its customers, including a "fixed" margin of profit, are directly tied to the value of its plant and equipment. It is this quirk in utility financing, among other things, which has driven the industry not only to dodge solar but to accept the risks and public relations headaches of a frontal assault on behalf of nuclear-electric plants. "The power companies' present drive for expansion," writes Sheldon Novick in *Environment* magazine (November, 1975), "despite all of the evident problems this entails, is not at all unreasonable from the companies' point of view. The easiest, and almost the only, way for power companies to increase the amount of money they make is to increase their capital investment and reduce their operating costs. Nuclear plants, which cost a great deal to build but which (theoretically) can be fueled more cheaply than coal or oil burning plants, do precisely this."

It is not without cause, therefore, that the gas and electric utilities have greeted the sudden appearance of solar energy with panic, hostility, and cooptive paternalism. Solar power has a built-in appeal to Americans of many stripes and political persuasions, from radical decentralists and environmentalists to family farmers and Republican suburbanites disgusted with ever-higher fuel bills and the sense of being victimized by faceless monopolies over which they have no control. "In an electrical world," writes Amory Lovins, "your lifeline comes not from an understandable neighborhood technology run by people you know who are at your own social level, but rather from an alien, remote, and perhaps humiliatingly uncontrollable technology run by a faraway, bureaucratized, technical elite who have probably never heard of you." Solar power inherently offers the prospect of liberation from the "uncontrollable technology" of centralized energy institutions. It is not so complex, in most of its most useful applications, that it can't be managed by persons other than a "technical elite." And it is "democratic," writes *Science* magazine: "It falls on everyone and can be put to use by individuals and small groups of people." It is also cost-efficient. In the five quick years of its U.S. renaissance, despite attempts to suppress, underfund, and slander its potential, solar energy has become not only a rival of natural gas and electricity for space and water heating in many parts of the country, it has proven itself a genuine threat to other uses of those expensive fossil fuels, including uranium, on the basis of which the utility industry has built its monopoly empire and mortgaged its future.

The utilities, of course, are not exactly helpless in the face of this looming competition. They have, for one thing, long enjoyed that form of collusion with the federal government exemplified by gas executive Henry Linden's seat on ERDA's General Advisory Committee. And recently, in conjunction with other steps to consolidate their defenses against such slings and arrows as solar power, they have deepened and broadened their influence within the federal energy establishment. Both EPRI and the American Gas Association, for example, have now signed what appears to be a legally questionable "Memorandum of Understanding" (MOU) with ERDA (DOE). It is legally questionable because the official government "Definition and Explanation of MOU" refers to "a written agreement entered into by the Administrator or Deputy Administrator of ERDA (DOE) and the head or his designee of *another Federal agency* (emphasis added)....The MOU establishes basic policy guidelines and a mechanism for coordinating cooperative activities." Nowhere in the official definition is there a reference to MOU's between ERDA (DOE) and private organizations such as EPRI and AGA. Furthermore, a list of ERDA (DOE) MOU's accompanying the "Official Definition" cites only federal agencies as partners to such agreements. (One of those partners, incidentally, is the Small Business Administration, but in the column marked "Program Coordinator," the line for SBA reads "None specified.") The list is dated June 2, 1977.

If it is not illegal, the "research and development" compact signed on May 25, 1976 by ERDA and EPRI certainly ought to be, for it is nothing less than an admission of EPRI—and thus the entire executive hierarchy of America's electric utilities— into a privileged sanctum of government where fundamental policy is determined, priorities established, and programs administered which bear explicitly on the lives, jobs, income, and even the health of every citizen in the United States (not to mention the children of those citizens who inherit the opportunity, say, to experience radiation poisoning from nuclear reactor melt-downs).

In fact, the sensitive nature of the ERDA/EPRI "Memorandum," which has clearly been used as a foil against small-scale solar energy, was disclosed by a nervous ERDA official in a telephone interview for this study. Such agreements, he said, "are usually signed between one government agency and another. The MOU with EPRI is a special case. It goes into more detail...discusses patent rights and procedural matters." The man was loathe to mail a copy of the document, even though its major provisions had already been publicized in EPRI's glossy house magazine. "The *EPRI Journal*," he said, "overstated the case on that MOU." But later he added: "Some of us (in ERDA) don't like the idea of signing MOU's with private organizations outside the government. We would prefer regular contracts." And still later: "I'd hate to see this MOU come out in a book. You (the author) might hold it out as a good thing—'Look, isn't this great?"—and somebody else would see it and complain. We don't want to look like partners with private enterprise. Congress just wouldn't stand for that."

The fellow can hardly be blamed for his queasiness. EPRI splashed the news of its MOU with the government across two pages of the *EPRI Journal* (July/August, 1976), treating the document as a "milestone" in the organization's history, le grand coup for the utilities at whose behest EPRI toils. "The two organizations (ERDA and EPRI)," crows the *Journal*, "are cooperating on 31 energy research projects with a combined funding value of approximately $84 million. Contemplated EPRI-ERDA projects would bring this figure to over $200 million. To formalize this productive working relationship, Dr. Robert Seamans, Administrator of ERDA, and Dr. (Chauncey) Starr (of EPRI) recently signed a Memorandum of Understanding that provides for broad energy R&D cooperation between the two organizations." Dr. Starr is quoted as saying that "we cannot overemphasize the importance of this agreement as a milestone in coordination between major sectors of society in their attack against the nation's energy problems." It is noted in the article that Dr. Starr, on November 27, 1973, had testified "before the House Committee on Government Operations on the proposed formation of the Energy Research and Development Administration (ERDA)," informing the committee that "we look forward to the opportunity of working in partnership with the proposed federal agency. There is a balanced set of roles which an agency such as ERDA can fulfill in the national interest and which EPRI will complement by helping the electric utility industry deliver energy to the consumer."

In describing the terms of the MOU, EPRI's magazine refers to "Information exchange," "Work for EPRI in ERDA facilities," "EPRI-ERDA joint funding of work by third parties," "Coordinated parallel or sequential contracting by ERDA-EPRI for related work by third parties," and "Such other activities, including stationing of personnel, as may be decided from time to time." In addition, "operations groups will be established for each discrete area, or group of areas, of common technical interest to achieve a parity of representation throughout the breadth of the relationship." EPRI and ERDA will also "provide for senior management overview of the cooperative activities undertaken," and toward that end, says the magazine, quoting from the MOU, "'the president of EPRI and administrator of ERDA, or their designees, will meet at least annually to exchange information concerning long-range programs, specific programs for the next fiscal year, status and/or results of all significant interactions, general progress and problems, and a summary of long-range and short-range programs of mutual interest.'"

While these citations reflect the broad scope of the EPRI/ ERDA "Memorandum," they actually understate the enormous advantage which the pact confers on EPRI as a "competitor" in what is supposed to be a "free enterprise" market for energy products and services, as well as federal R&D contracts. Not only is EPRI awarded a role in deciding what categories of energy R&D the government will fund, it is guaranteed untold millions of dollars in contracts assigned on the basis of those decisions. EPRI is further entitled by the 15-page MOU to be kept abreast of ERDA contract work which may concern EPRI not as a funding partner but as a "parallel" contractor on another job: "Each party will endeavor to secure for the other (non-funding) party the right to visit contractor work sites for purposes of technical evaluation at reasonable intervals and upon reasonable notices." In other words, if Solar Dynamics, Ltd., of Austin, Texas happens to be working on an ERDA-funded project in the same general category as one of EPRI's operations, EPRI has the right to "evaluate" that project.

The "patent and data" clause of the MOU, though scarcely comprehensible, appears to be as generous to EPRI as the rest of the "Memorandum." It grants the organization and its "member utilities," both at home and abroad, "an irrevocable, non-exclusive, royalty-free license...to make, use, and sell...any invention or discovery made or conceived in the course of or under jointly funded efforts and covered by a U.S. Patent." Still another gift to EPRI is a stipulation that when ERDA performs work "for EPRI in an ERDA facility, ERDA will normally establish, at a reduced rate for the work in accordance with applicable ERDA policy, the same rates it usually charges to federal agencies, i.e., rates that exclude depreciation and the 'added factor'." This is followed by a comic admonition that both parties attempt to "avoid conflicts, or appearance of conflicts, of interest in their jointly sponsored activities and contractual and subcontractual relationships."

Naturally, the government's wedding with the utilities much transcends the paper and ink of the nuptial agreement. It is enhanced by a dowry of billions of the taxpayers' dollars (to be discussed momentarily), plus a warm convergence of habits, opinions, values, and a circle of mutual friends. Joseph Fisher, for example, a Democratic congressman from Virginia now in his third term, is as follows: (1) a member of the powerful Ways and Means Committee of the House of Representatives, with substantial jurisdiction over federal energy concerns; (2) a member and founding chairman (in 1973) of EPRI's blue-ribbon Advisory Council; (3) a former president of Resources for the Future, having resigned in 1974 to run for Congress. "There remains no doubt in Congressman Fisher's mind," says an adulatory article in the *EPRI Journal* (April 1977), "that whatever our energy future, EPRI has a major role to play. He says that government agencies will be looking to Congress, as well as to EPRI and private industry, for guidance and advice, especially in the year of government change (i.e., the transition to the Carter administration and establishment of the Department of Energy)." It was Fisher who, in league with House Speaker Thomas P. O' Neill and other congressional leaders, put together in May of 1977 the House Ad Hoc Committee on Energy which lobbied so effectively in the House for the Carter-Schlesinger "Energy Plan."

Not surprisingly, Fisher's views on America's resource priorities match EPRI's to a tee. "With a background in economics and a reputation as a national energy leader," says the *EPRI Journal*, "Fisher feels a special responsibility to warn the public not to be taken in by their 'love affair with solar energy.' He does not hesitate to caution his fellow congressmen to 'watch themselves and not overrespond to solar or any other new technology with more money than there are ideas or good people to develop it.' Fisher sees the enthusiasm over solar energy, for all its more distant promise, as an escape, for many people, from other problems." Meanwhile, says the magazine, Fisher believes that the nation must "face the 'dirty, mucky problems of how to use more coal without too much damage to the environment and health.' He also says there is no choice but to increase our reliance on safe nuclear power." (Fisher's intimation that America must simply buck up to the hazards of a "hard-path" energy future typifies a corporate cynicism, indeed a fatalism which was chillingly expressed in the Ford Foundation's 1977 report on *Nuclear Power, Issues and Choices*, cited earlier in this study. The Ford document, according to *Energy Daily* [March 22, 1977], while not glossing over the dangers of nuclear power plants, does argue that "the consequences of a major disaster 'would not be out of line with other peacetime disasters that our society has been able to meet without long-term social impact.'")

The fraternal character of EPRI's relationship not only with the federal government but with other sectors of the corporate elite was further underscored in a panel discussion printed in the *EPRI Journal* of January/February, 1978. Among those participating in the forum, moderated by *Energy Daily* editor Llewellyn King, were most of EPRI's senior division managers and president Chauncey Starr. It was revealed by the panelists that EPRI's origins date back to 1971, when the Electric Research Council (EPRI's predecessor) of the Edison Electric Institute released a study calling for an accelerated R&D effort by or at least on behalf of the utility industry. "There was at the same time," said EPRI's Rene Males, formerly of Commonwealth Edison in Chicago, "a drive to get the federal government into the picture. The industry felt quite strongly that it was important to play a role in order to control their (the utilities) future." Males also mentioned "the need to be sure and stay on top and complement federal government R & D work." This is accomplished, said Rudman, who left IBM to become director of planning for EPRI, in the following manner: "We work very closely with the key government agencies. We always send them copies of our program plan, and our program managers work with them on a weekly basis. So we are aware of what is going on there and they are aware of what is going on within EPRI."

Talk at the forum inevitably turned to the breeder reactor, and particularly to Jimmy Carter's announced intention of halting work at the Clinch River site. "It is not at all my perception," said Milton Levenson, EPRI's nuclear manager, "that it is the government's policy not to proceed with the breeder." Chauncey Starr added that there is "no way this nation can meet its energy needs in the next 25 years without expanding nuclear, and there is no way to expand nuclear without having the fast breeder come through. So I think all the president can do is delay. I don't think he can prevent the eventual denouement of the breeder." Starr also took the opportunity to call solar energy a "miracle" whose adherents refuse to believe "does not exist," despite EPRI's efforts to convince them. "They have to believe that there is a miracle alternative," said Starr, "and solar is that one." As for EPRI's role: "We don't have a responsibility to promote solar heating. We do have a responsibility to thoroughly analyze what the development of solar heating and cooling would mean to the electric utility systems and their customers in the United States and to the total energy picture." (Why, one must ask, is it necessary to analyze the potential effect of a "miracle" that "does not exist"?)

The panelists agreed that EPRI is not so much an "innovator" in the field of energy R&D as it is a "catalyst" for the R&D programs of a variety of interested parties, including major oil companies. Richard Balzhiser, director of EPRI's Advanced Systems Division, referred to "the area of coal liquefaction. We've developed a group of people who are technically very capable. By virtue of our orientation outside the petroleum industry—an industry that because of anti-trust and other things has virtually no interaction (on a formal basis with the utility sector)—we find this industry coming to us with their ideas. Our group has become a focal point for interaction among the scientists." Ric Rudman noted that EPRI had "a couple of major projects right now: at Exxon we have $40 million, at Powerton we have $30 million."

The existence of such programs in EPRI's portfolio, not to mention the MOU with the Department of Energy, suggests the likelihood of an executive revolving door, which was precisely the point of a comment by Rene Males. EPRI, he said, "is where the action is. This is really the leading edge of research. These people will probably stay with EPRI for three to ten years, develop their skills and their programs, and then go back into a university setting, to a federal agency, or to a utility or manufacturer."

There could be no clearer example of this revolving door than the fact that Floyd Culler, deputy director of Oak Ridge National Laboratory, was selected to replace Chauncey Starr as president of EPRI when Starr retired to a less visible role on June 1, 1978. The "urbane and dynamic" Culler, as *Energy Daily* portrayed him, moved to Palo Alto in January, 1978, to commence his EPRI apprenticeship as executive vice-president. It was equally suggestive—and prophetic—that three weeks before the signing of the EPRI/ERDA (DOE) "Memorandum," on May 5, 1976, Chauncey Starr and Robert Seamans, Administrator of ERDA, showed up together as the U.S. delegates to an International Symposium on Electricity Research and Development in Washington, D.C. There, according to the *EPRI Journal* (January/February 1977), the two officials "summarized national energy policy issues" and joined with "senior policy-makers" from Britain, France, Japan, and Sweden in "a broad survey of the research and development planning currently underway" in the nations represented at the symposium. EPRI, in addition, is the permanent U.S. representative to the International Electric Research Exchange, which meets at least annually to facilitate policy and research coordination between the utility industries of 14 "non-communist" countries.

These examples merely scratch the surface of EPRI's many international responsibilities, including policy discussions ordinarily conducted by governments, and the importance of such a role in defining America's energy direction can hardly be exaggerated. Recent studies of the subject—most notably *Global Reach* by Richard Barnet and Ronald Müller—have made it clear that the U.S. corporate elite now views itself and behaves routinely as the leading element in an international corporate order whose operations frequently transcend the sovereignty of national governments. This has required, as mentioned in previous chapters, certain adjustments in the machinery of coordination between the corporate and government sectors not only of the United States but of other nations, especially those in the "trilateral" economic sphere. "The power of the global corporation," write Barnet and Müller, "derives from its unique capacity to use finance, technology, and advanced marketing skills to integrate production on a worldwide scale and thus to realize the ancient capitalist dream of One Great Market. This cosmopolitan vision stands as a direct challenge to traditional nationalism."

The authors quote Jacques Maisonrouge, chairman of IBM, to the effect that the "'world's political structures are completely obsolete. They have not changed in at least a hundred years and are woefully out of tune with technological progress.' The 'critical issue of our time,' says Maisonrouge, is the 'conceptual conflict between the search for global optimization of resources and the independence of nation-states.' " Meanwhile, in his book *Between Two Ages* (1970), "trilateralist" Zbigniew Brzezinski (now Jimmy Carter's foreign policy adviser) has called for "a new pattern" in national and international government structures, "a blurring of distinctions between public and private bodies." Brzezinski proposes, among other things, "a world information grid, for which Japan, Western Europe, and the United States are most suited," as well as "a more rational division of labor in research and development" of mutual interest to the trilateral nations."

Where better to advance a "blurring of distinctions between public and private bodies" than the utility industry, which has blurred such distinctions since its early days as a "regulated" monopoly? And what better man than Chauncey Starr—the articulate former president of the Atomics International division of Rockwell International—to help the government devise and implement an acceptable international strategy for electrifying the industrialized west? Here we return to the larger corporate energy scenario discussed in Chapter I, among whose objectives is reduced dependence on foreign petroleum through an expansion of coal and nuclear-fired electrical plants. Since Starr has functioned as a principal author of that policy, it was natural for him to sketch its basic tenets before an audience of executive engineers at a 1976 "Energy Awareness" symposium in Knoxville, Tennessee. (A revealing account of that symposium is presented in the following chapter.)

Starr's address, entitled "A Strategy for Electric Power," was peppered with statistics and projections exactly matching and in some cases drawn from studies by ERDA (DOE), Chase-Manhattan Bank, and the Ford Foundation. He assumed, for example, that America would require 170 quads of energy per year by 2000 (see *ERDA 49* and *Exxon USA*, First Quarter, 1976). He then told his audience that in order to obtain such a quantum of energy and still "keep oil import levels about what they are now (7 million barrels per day)," the U.S. electrical grid will have to be enlarged to provide a staggering 63 percent of the nation's total energy supply. Of that amount, he said, coal will generate 47 percent and nuclear power another 47 percent, with hydro-electric plants accounting for most of the rest. This corresponded with a subsequent projection by Dr. Richard Roberts, director of ERDA's nuclear program, who said that "nuclear energy this year will provide about 9 percent of our electrical energy, will increase to about 20 percent in 1985, and will reach as much as 50 percent by the year 2000."

Such a dovetailing of figures and projections, which occurred repeatedly throughout the Knoxville symposium, brought one of the lesser government speakers to observe, perhaps in awe, that "I think there is a very clear strategy that has been set out this morning on the part of government and industry." Earlier in the program, when Dr. Herman Postma of Oak Ridge National Laboratory was asked about the implications of that strategy with regard to American foreign policy, he replied: "It is my belief that the strategy that has evolved for the energy future of this country has a fundamental international aspect to it." Indeed: "I would conclude that international considerations were not only considered in the evolution but dominated the entire strategy."

The veritable centerpiece of that strategy, as evidenced by Secretary Schlesinger's visit to Brussels in October, 1977, is development of the liquid metal fast breeder reactor, a "miracle" device which promises to deliver the international capitalist economy from the clutches of the Arabs by producing more fuel than it consumes. Unfortunately, however, the U.S. breeder program has been delayed by technical problems, citizen opposition, and government fears concerning the threat of nuclear weapons proliferation. (The breeder, unlike conventional light water reactors, is fueled by plutonium, a few pounds of which will make "dandy" atomic bomb.) As a measure of corporate impatience with these delays, the Rockefeller Foundation released a study in May 1978 calling for a "bilateral U.S.-Japanese program" in which "three breeder demonstration plants (would) be developed, designed, and constructed—two in the U.S. and one in Japan." According to a report on this study in *Energy Daily* (May 15, 1978), such a program would serve "not only as an 'insurance policy' against future energy shortages but also as a means for the U.S. to move back into international leadership on nuclear matters." One month later, a parallel study was issued by the Trilateral Commission (founded by David Rockefeller) in Washington, D.C., this piece co-authored by none other than John Sawhill, the New York "trilateralist" who directed the FEA under Nixon. "Sawhill," wrote *Energy Daily* (June 16, 1978), "called on trilateral governments (via a Washington press conference) to develop a joint policy on nuclear energy and nuclear weapons proliferation in light of the recent conflicts between the U.S., Japan, and Europe over breeder reactors."

It hardly seems an accident of circumstance that just three months before these studies were released, EPRI announced a "breakthrough" in the development of a "proliferation-resistant" fuel cycle for the breeder (called the Civex process) on which it had been working with the British Atomic Energy Agency. And lo, on that very day, according to an *Energy Daily* bulletin (March 2, 1978), "Chauncey Starr, the 'grand old man' of U.S. electric power research," was awarded the highly prestigious Legion d' Honneur by the government of France. "There is a certain fine irony to the honor," said *Energy Daily*, since earlier that day Starr had "unveiled to the world a proliferation-resistant nuclear fuel cycle technology that would advance the cause of the breeder reactor—in which France currently has a head start."

Amidst the pomp and ceremony of Starr's award (he was kissed on both cheeks by the French ambassador), it was not widely acknowledged that EPRI and DOE/ERDA have been quietly pursuing a joint program of breeder R&D for at least four years. In fact, the plutonium breeder has been the object of more joint funding and research by EPRI and the U.S. government than any other single technology, revealing still further the import and purport of the EPRI/ERDA (DOE) "Memorandum of Understanding." In 1971, says *Energy Futures* (an ERDA-funded industrial survey mentioned earlier), the federal government, "responding to growing industry and AEC enthusiasm for the breeder...declared the technology to be its most important priority." The survey observes that the three corporations most heavily involved in breeder R&D—Westinghouse, General Electric, and Rockwell International—had until 1971 "paid for most of their own work, investing more than $30 million apiece by the late sixties. Westinghouse soon gained a pre-eminent position in the early seventies, and it has now spent $40 million. But today the government pays for 80 percent or more of the work done by all three companies." Thus, says the survey, "ERDA's (breeder) budget in fiscal 1976 was $428 million, and will be $575 million in fiscal 1977—one-fourth of ERDA's entire research, development, and demonstration budget." (DOE breeder budgets for 1979 and 1980 are $754 million and $611 million respectively.)

EPRI entered the picture in 1975, says *Energy Futures*, joining ERDA "to sponsor design studies for the 1000-MW commercial prototype. (ERDA) expects the utilities will build the plant by the late eighties, with ERDA contributing somewhat less than half the cost. Westinghouse, G.E., and Rockwell each contracted to perform competitive $10-million designs." The survey adds that although "ERDA has stated officially that it will not decide until at least 1986 whether to pursue commercialization of the LMFBR, it nevertheless has obtained pledges from the three companies (to) design and build 1000-MW powerplants for utilities after 1978." Westinghouse, for its part, "intends to begin marketing its design that year, and hopes to sell ten per year thereafter." All three companies "believe that continuing financial support from the government is vital. 'Industry can't carry it on its own if we're going to keep the competitive system,' says William V. Potts of Rockwell." Not to worry, says *Energy Futures*: "The massive program is pushing ahead. ERDA intends that some reactors be built, and it is accelerating its efforts." While Britain and France, recalling the international dimension, "have successfully operated demonstration breeder-type reactors," enough "money and manpower have been committed to the American program that it has powerful momentum."

* * * *

The above discussion of "trilateral" corporate interest in breeder reactor technology is intended primarily to illustrate the role of America's utilities—represented by EPRI—in a broader corporate mission to salvage the capitalist system itself through a "hard-path," capital-intensive, centralized energy future. Of equal significance in the corporate/government "game-plan" are such technologies as conventional nuclear power plants and "advanced" fossil fuel technologies, particularly coal. A discussion of the latter will follow momentarily, that followed by a glimpse at the anti-solar consequences of the huge federal investment in these technologies. Meanwhile, it is worth pausing to examine the more immediate pecuniary interests of the corporate elite in a "trilateral" nuclear power industry.

The political force of such interests was highlighted early in 1978 by evidence of an ugly quarrel in Washington between the Council on Environmental Quality (CEQ) and the Nuclear Regulatory Commission (NRC). According to an article on January 19 by columnists Rowland Evans and Robert Novak, "an angry counterattack against federal environmentalists is now being quietly planned by Cabinet-level departments, led by the State Department, with indications of support in the White House itself." The furor resulted from a plan by the CEQ, apparently co-authored by the privately funded Natural Resources Defense Council, which would, according to Evans and Novak, "require standard environmental impact statements (to be called 'assessments' in the foreign field) for all exported material or technology sold abroad with some help—export licenses or loan guarantees—from the U.S. government."

This would be "outrageous," said a State Department official quoted by the columnists: "These regulations would impose American environmental standards on all our foreign friends and they would end up hating us." The "real target," say Evans and Novak, "may be nuclear reactors, a prospect that has infuriated the NRC." The columnists add that at "least as upset as NRC and the State Department (is) Export-Import Bank president John Moore (a former Atlanta lawyer and friend of President Carter), who warned...that the proposed regulations would benefit Japanese and West German (nuclear) exporters at the expense of this country, (since) endless delays and lawsuits against U.S. exports on often specious environmental grounds would turn impatient foreign buyers away from the U.S."

A hidden reality behind this "outrage" at the CEQ's attempts to regulate nuclear exports by U.S. companies is the simple fact that without such exports the nuclear divisions of Westinghouse and General Electric—the nation's major producers—would apparently go bankrupt almost overnight. Both corporations have been damaged severely by a sharp drop in orders from American utilities—only two reactors were ordered in 1977 and none in 1978—as well as by delays and costly malfunctions in perhaps the majority of their other nuclear plants. So acute is the problem that General Electric has informed the Carter Administration, according to a speech by Barry Commoner on February 4, 1978, that if something isn't done to stimulate sales of nuclear hardware the company will shut down its nuclear operation altogether. It is quite possible, of course, that the message was delivered by G.E. board chairman Reginald E. Jones, who occupies the co-chair of the "President's Labor-Management Group." At any rate, on Carter's trip abroad in January, 1978, he accepted orders for eight nuclear reactors from the Shah of Iran and two from the president of Brazil.

Meanwhile, corporate-government fears pertaining to "endless delays and lawsuits" against U.S. nuclear exports have recently been substantiated by accounts of mass-resistance to atomic power plants in France, West Germany, and other nations. "The fact is," writes Amory Lovins in *Foreign Affairs*, "that in almost all countries the domestic political base to support nuclear power is not solid but shaky. However great their nuclear ambitions, other countries must still borrow that political support from the United States. Few are succeeding. Nuclear expansion is all but halted by grass-roots opposition in Japan and the Netherlands; has been impeded in West Germany, France, Switzerland, Italy, and Austria; has been slowed and may soon be stopped in Sweden; has been rejected in Norway and (so far) Australia and New Zealand (and) two Canadian provinces." Then, following an impudent suggestion that the U.S. "phase out its nuclear programs," Lovins writes that his own judgment, "based on the past nine years' residence in the midst of the European nuclear debate (Lovins resides in England), is that nuclear power could not flourish there if the United States did not want it to."

It will now be recalled that nuclear power, for all its promise in the desperate minds of those who subscribe to it, represents only 50 percent of the electrical energy future proposed by EPRI/DOE and their "hard-path" corporate sponsors. The other 50 percent is represented by coal. Hence a 1977 ERDA report which concludes that while the breeder reactor continues to enjoy a long-term "economic edge over conventional and most competing new technologies," advanced coal technologies must be pursued as well—to the detriment of solar power. "Judging from its tenor," writes *Energy Daily* (January 7, 1977), "ERDA's *Comparing New Technologies for the Electric Utilities* is likely to disappoint solar thermal power buffs and bring joy to those who are pushing advanced coal-burning techniques such as fuel cells, MHD, fluidized bed combustion, and combined cycle turbines with gasifiers—all of which ERDA considers of great economic promise." (The ERDA executive in charge of writing that report, incidentally, was Fred Weinhold, who represented the Ford Foundation on two different panels of Dixie Lee Ray's energy R&D study for Nixon in 1973.) These new coal technologies are so important that the January/February, 1977 issue of *ERPI Journal,* in a summary of its R&D activities, devotes more space to coal than to any other resource, calling it "The R&D Pivot for New Energy Fuels."

A conspicuous feature of the EPRI/DOE coal program, as of the nuclear program, is the role it assigns to such venerable corporations as Westinghouse and Rockwell International (Chauncey Starr's former employer). In one recent project, for example, EPRI and DOE brought "four major oil companies" into the design and construction of a pilot plant for coal liquifaction ("H-Coal") in Catlettsburg, Kentucky, completed in 1977. Another project by EPRI and DOE, this one including General Electric, is a "fluidized-bed combustion system" for coal-fired plants which should, according to the *EPRI Journal*, reach "full-scale commercial development in the early 1980's." Combustion Engineering, Inc. teamed up with EPRI/ERDA to complete in 1977 a pilot model "two-stage entrained gasifier" at Windsor, Connecticut, and United Technologies, Inc. has been funded by DOE to fabricate a 4.8-megawatt "fuel-cell demonstration" plant in 1979. "This important ERDA (DOE) commitment," says the *Journal*, "is a direct result of utility industry and EPRI involvement in fuel cells."

That sort of boasting likewise accompanies a discussion in the magazine of an elaborate coal-gas process called "magneto-hydrodynamics" (MHD). Evidently the MHD program had been troubled by government safety regulations, so EPRI, "by taking a comprehensive view and then focusing on the optimal generator design, helped to relax design requirements throughout the system. This should significantly influence the ERDA (DOE) design philosophy and lead to a more flexible, comprehensively designed plant." The magazine further observes that "EPRI has a small but significant program in coal-fired, open-cycle MHD technology in order to provide maximum utility industry input at minimum funding to a large ERDA (DOE) program." (This article was written by Richard Balzhiser, the former Nixon adviser who left the White House Office of Science and Technology in 1973 to become EPRI's Director of Fossil Fuels and Advanced Systems, as well as a member of ERDA's Fossil Fuels Advisory Committee.) A sampling of other partners with ERPI and DOE in their pursuit of new technologies for coal includes the Battelle Memorial Institute, Rockwell International, Arthur D. Little, Inc., Union Carbide, Babcock and Wilcox, Exxon Research and Engineering, Westinghouse, Texaco, Occidental Petroleum, and Mobil Research and Development Corporation. The latter firm, according to Balzhiser, "was recently awarded the Bituminous Coal Research Award (by the National Coal Association) for its innovative work for ERDA."

The gas utilities, represented by the American Gas Association, are proceeding in a similar vein with what the Association's magazine calls an "ERDA/AGA Coal Gasification Pilot Program." America's "supply of natural gas," says the magazine, "continues to be critical. The need to supplement this gas with pipeline quality gas from coal is even more critical today than when the Joint Program started (in late 1971)." The object of the $150-million program, the government's share of which comes to $100 million, is to bring some four or five techniques for coal gasification to the pilot plant stage by early 1978.

This and the parallel DOE projects shared by EPRI and its corporate companions bear out clearly a remark by Robert Engler in *The Brotherhood of Oil*: "the (oil and gas) industry envisions a tremendous expansion of electricity generated by coal-fired steam turbines and a network of huge gasification plants for the transmission of gas eastward." Many of these plants, however, due to the gross toxicity of their operation—involving fierce temperatures and violent chemical reactions—will introduce a whole new range of ecological threats. In a coal-scrubbing process known as "flue-gas desulfurization (FGD)," for example, a calcium sulfate "sludge" removed from smoke stacks is presently "stored" in ponds of water adjacent to the plants. From these ponds, according to the *EPRI Journal*, "salts and toxic trace metals may, under certain conditions, leach into the groundwater. The pond problem could become massive. Between 1976 and 1990, utilities will invest about $30 billion for 200,000 MW of FGD systems. This implies that over the next 15 years an estimated 1.5 billion tons of sludge will be accumulated—enough to cover 50,000 acres to a depth of 20 feet." Such are the beauties of high technology, and hence Engler adds: "Industry and government have often been quite secretive as to where these plants are to be located. 'For competitive reasons' is the usual explanation. They obviously fear adverse community reaction and the demand for public hearings. A 'highly confidential' study by the AGA located 176 sites for commercially feasible coal gasification plants."

As might be inferred from the figures for the ERDA/AGA gasification program, the costs of the government's coal and "synfuels" R&D are imposing. Dixie Lee Ray's calculations for Nixon suggested a five-year appropriation of $2.2 billion for such R&D, and the 1978 DOE budget in fact allotted it almost $600 million, the preponderance being shared by EPRI, its coal and gas industry counterparts, individual utilities, and "energy" corporations. That $600 million represented 11 percent of DOE's total R&D expenditures for 1978—and the figure shoots skyward when it is added to the federal investment in nuclear R&D, especially after compensating for an apparent idiosyncracy in DOE's bookkeeping procedures. If, for example, one doesn't count such items in the FY 1978 nuclear R&D budget as "Uranium Enrichment" and "High Energy Physics," which were mysteriously not listed as "Research, Development, and Demonstration" costs, the appropriation for nuclear power was $1.9 billion, including $656 million for the breeder reactor. If, on the other hand, one does count the former items in the nuclear R&D column—and there are persuasive reasons for doing so—the nuclear commitment in 1978 was $3.1 billion, or 60 percent of DOE's entire budget "revised" to include the otherwise missing items.

Added to the coal and synfuels appropriation, this brought the federal investment in utility-related R&D for 1978 to roughly 70 percent of DOE's total applicable budget of $5.2 billion. The solar share of that budget (ignoring for a moment the fact that most of those funds, too, reach the corporate elite) was $305 million, or 5.8 percent, and presto—here is the strongest single weapon in the utilities' arsenal against a rapid development of solar technologies. (Indeed, since 1918, according to a study in 1978 by Battelle Pacific Northwest Laboratory, the federal government has spent *at least* $133 billion on "incentives" for the oil, gas, coal, and nuclear industries, versus less than $1 billion for solar.) The significance of such a weapon, moreover, reflecting the welter of interlocks between the utilities and the government so clearly manifested in the EPRI/ERDA (DOE) "Memorandum of Understanding," extends far beyond the billions of dollars involved. Amory Lovins, as usual, puts it well: "The money and talent invested in an electrical program tend to give it disproportionate influence in the counsels of government, often directly through staff-swapping between policy and mission-oriented agencies. This incestuous position, now well developed in most industrial countries, distorts both social and energy priorities in a lasting way that resists political remedy."

IX

UTILITIES AND SOLAR: DECLARING WAR ON THE COUNTERCULTURE

In further tracing the interweave between America's utilities and government energy planners, it must be kept in mind that the threat they observe in small-scale solar energy derives in part from their identification of the popular "solar movement" with those urchins in the "counterculture" who are trying to stop the expansion of nuclear power—and perhaps to foment social revolution. Due to long and often bitter experience with this "counterculture," there are many in the U.S. energy establishment who now consider it imperative to thwart the anti-nuclear lawsuits and other activities of citizens' groups like Friends of the Earth, Critical Mass, and New England's Clamshell Alliance. (The latter group in 1977 and 1978 staged massive sit-in demonstrations aimed at blocking construction of the Seabrook nuclear power plant in New Hampshire, setting an example of direct non-violent action which has since been emulated by other anti-nuclear "alliances" from Oregon to Oklahoma). This equation in the corporate mind between getting on with the "hard path" and blocking the "counterculture" was dramatically and instructively revealed at a February, 1976 "Energy Awareness" seminar in Knoxville, Tennessee, where Chauncey Starr of EPRI mingled with other top executives from the corporate-government elite to discuss mutual energy goals and a body of tactics for achieving them.

Knoxville lies adjacent to the government's Oak Ridge National Laboratory, the nuclear capital of the world, and the "Energy Awareness" symposium, according to a transcript of its proceedings, was logically sponsored by a coalition of 16 nuclear-related engineering societies "in cooperation with: Americans for Energy Independence, the U.S. Department of Commerce, FEA, ERDA, and the Union Carbide Corporation (which has long played a central role in the operation of the Oak Ridge labs). The ostensible purpose of the meeting, as stated by a Union Carbide executive who helped organize it, was "to urge the leadership and representatives of prominent civic, fraternal, service, labor, trade, educational, and technical organizations to establish programs for the dissemination of factual information on national energy issues." What the speaker meant by "national energy issues" was nuclear-fired electricity, and the real purpose of the meeting was to inspire those present to join the utilities and the government in a Great Crusade not merely on behalf of nuclear power but against such scoundrels as Ralph Nader and his "counterculture" legions who would spoil the greatness of America by tripping its leap toward an "all-electric economy," as Dr. Starr phrased it, in the year 2000.

The keynote speaker at an evening banquet was U.S. Representative Mike McCormack, himself a nuclear engineer once employed at the Hanford facility in Washington, now a member of the House Science and Technology Committee and chairman of a special subcommittee on the federal breeder reactor program. Following his introduction as "the leading Congressional advocate of nuclear fission" who "has obtained substantial increases in (nuclear) funding for 1974, 1975, and 1976," McCormack addressed the major themes and issues of the "Energy Symposium" with a vengeance. Citing projections by the Ford Foundation's Energy Policy Project, he declared that even a reduced energy growth rate will "double our energy consumption within about 25 years...and that means doubling electric energy production by 1990." Since domestic reserves of petroleum are dwindling, he said, and "the situation in the Middle East remains unstable," the nation "will become dependent on coal and nuclear fission." These options represent "our only hope of energy self-sufficiency and economic stability during the balance of this century," and "like it or not, the public must be made aware that there is no choice." He added that if the 170 nuclear plants now on order by utilities are completed "by 1985—and they can be if we simply eliminate unnecessary delays and provide for construction capital—then this nation will have a nuclear capacity of about 225,000 megawatts, or about 30 percent of our electricity-generating capacity in 1985 (compared with 8.5 percent in 1976)." Looking toward the year 2000, McCormack promised additional funding for the plutonium fast-breeder and said that "if we assume the nuclear program currently assumed by ERDA and my subcommittee on the breeder study...we would wind up with 600 to 800 nuclear power plants on line" by century's end. Meanwhile, he said, "even a tidal wave of federal funding" could not "make solar or geothermal energy a significant resource before the year 1990, and it would be very small even then."

McCormack ridiculed opponents of nuclear power who fear the hazards of reactor core defaults and radioactive emissions and wastes. "Overdoses of aspirin compounds," he said, "cause hundreds of deaths each year. About 1000 persons die from electrical shock," and "more than 2000 are bitten by rabid animals. I hold it is more dangerous to own a dog than to live in a nuclear power plant." Concluding that nuclear energy on a grand scale is necessary for everything from jobs and a clean environment to "women's liberation," McCormack appealed to his audience "to help provide public awareness of the problems that are facing us. Like it or not," he said, "all of us in this room...have been drafted." Just as "many of us were drafted to serve our country in World War II, today we have been drafted by fate because we care about our country." The most pressing objective, he told his new cadres, was to get out and hurl back the "anti-nuclear activists" who were trying to stop nuclear power through a series of "nuclear moratorium" referenda scheduled that year in a number of states—particularly the Proposition 15 election to be held June 8 in California. McCormack himself had made a recent speech against that initiative, and he urged that the battle be taken "to the labor groups and the business groups and the fraternal groups...If organized labor would simply do a thorough study on what the initiative means to them out there, it wouldn't have a prayer."

McCormack told his warriors they would also have to pay special attention to the press, which "has printed so much garbage on this subject over the last year. I wonder why it is more important to them to publish some feature story about why some farmer in Scotland can run a car on garbage than to write an instructive story about the energy policy in this country." There are, he said, "an awful lot of lazy reporters," as well as editors who "take whatever is given them; and when Ralph Nader puts out a press release, he puts it on the wires all over the country. Then the nonsense he creates pops up as truth in the newspapers because we are not doing anything about it." Another critical target for pro-nuclear forces, said McCormack, are state and national legislators who "have been brainwashed by the Friends of the Earth or some similar group." He said he didn't know how many of his House colleagues had been "brainwashed, but I do know key members of Congress who are constantly being visited by delegates from Friends of the Earth or Nader organizations before they go to committee meetings." In fact, he continued, certain brainwashed congressmen had tried just recently, without success, to make objectionable amendments to the Price-Anderson Act (a federal nuclear insurance program to compensate for the refusal of private insurance companies to issue policies on nuclear plants). These congressmen, said McCormack, had been "fed" the amendments by Nader zealots "as an antinuclear act. They had no idea what they were doing. They were just trying to goof up the operation." He closed his speech with a final call to arms, comparing the struggle ahead with the "Depression" and "World War II," but expressing confidence that "we can overcome the problems that face us today and help build a better world." As McCormack sat down, the Union Carbide man who had introduced him quickly restored a sense of perspective: "When you start handing out the billion-dollar programs, remember us, will you, Mike?"

* * * * *

Among the nuclear drum-beaters preceding McCormack's appearance were some who had been active against solar energy since 1972. Dixie Lee Ray addressed the symposium (in her capacity as Assistant Secretary of State), followed by EPRI's Chauncey Starr, Paul Turner of the Atomic Industrial Forum, Robert Price of the National Coal Association, and Dr. Wilson Laird, a former executive with the Interior Department now doing much the same work for the American Petroleum Institute. Also sharing the podium were three top officials from ERDA (including the Assistant Administrator for Nuclear Energy), an FEA official, the director of Oak Ridge Laboratory (who gave the token speil on solar power), the chairman of TVA, and a union representative named J.C. Turner, currently a member of what he called "the President's Labor-Management Committee." Another speaker and one of the organizers of the meeting was Endicott Peabody of the famous coal family, a former governor of Massachusetts and presently co-chairman (with nuclear physicist/Nobel laureate Hans Bethe) of Americans for Energy Independence, a "citizens' energy awareness" organization which Peabody said had a newsletter circulation of 30,000, plus a "speakers bureau" and a strong "educational program" aimed at the country's public schools.

While as many as half the speakers that day shared Mc-Cormack's crusading and quasi-paranoid demeanor, perhaps the most striking in that regard was Richard Myers, managing editor of the influential *Energy Daily*, who apparently sees Naderite subversives behind every solar collector. Myers portrayed the struggle against nuclear power as a struggle against the "growth" of urban industrial society in general. It is, he said, "a ground-swell of anti-business, anti-industry sentiment...a revolt against bigness and large centralized bureaucracies. Ralph Nader, for example, speaks of replacing nuclear power with solar power, which is a proposition I find wholly unreasonable." (If you buy a solar unit, said Myers, "I would advise you to get something fired with oil and gas to back it up.") Myers described the push for solar as a "push to decentralize, a push to break things down into self-sustaining units...that is truly motivating these no-growth advocates. I think the roots of all this go back a very long way...to some of our more recent national traumas, like Watergate and Vietnam. I think it reflects a general fear and distrust of the system, particularly a large system in which the public cannot perceive any individual to be held accountable. But I also think Ralph Nader and company are wrong on this point." Indeed, Myers compared "Nader and company" with a "mob," referring to a "twentieth-century philosopher who said that when the mob gets hungry for bread, the first thing they do is go out and wreck the bakery."

Myer's "bakery" is of course the nation's existing energy system, and rather than "wreck" it he wants to expand it with coal and nuclear power, and that, he said, will take "capital, financial resources, or, in short, money"—of which unfortunately there isn't enough available. He cited a Chase-Manhattan Bank study indicating a "capital shortage" by 1984 of "approximately $1.5 trillion," and then told his audience that "one of the major reasons (for the shortage) is that we have, in this country, institutionalized a great number of social programs. The American Enterprise Institute in Washington (whose 'National Energy Project' is chaired by former Secretary of Defense Melvin Laird) recently published a report that estimated there are over 100 federal programs transferring benefits to the poor." To fund these programs, said Myers, "the Federal government in the last 10 years...has taken 350 billion dollars out of the private capital market. That is 350 billion that was not available to private industry." (Myers failed to mention the trillion-plus dollars spent for "defense" in that period, including the Vietnam war under Secretary Laird.) Since "energy companies," he said, "are routinely committing a billion dollars to a single facility (such as nuclear and coal gasification plants)," and since "industry is not used to making investments of this order," it is obvious that the government will have to provide a little boost in the form of "loan guarantees" and collateral programs for industry. Myers seemed inclined, for starters, toward "an ambitious proposal, being promoted by Vice-President Nelson Rockefeller, to create a 100-billion-dollar energy bank, a government energy bank." (This motion was later seconded by none other than J. C. Turner, the president of the International Union of Operating Engineers, who remarked that he had "attended a breakfast two weeks ago with Governor Rockefeller in which he talked about the plan.")

Myers left it unclear whether he believed that America's "social programs" ought to be abandoned to release still further billions for nuclear power, but there was no mistaking what he thought should happen to Ralph Nader's "powerful coalition of local citizen action groups" who were "pushing for a moratorium on nuclear power development." Referring specifically to the Proposition 15 initiative coming up in California, Myers argued that the activists out there "are the same people, with the same ideology, who want to break up the large oil companies... who want to extend price controls on the oil and gas industry... who would like to restrict the coal industry's ability to mine coal and the electric utilities' ability to burn it...who tend to fight their local utilities' efforts to pass through (to the consumer) increased fuel costs...who oppose any kind of government financial support or incentive to any of the energy industries." Could the audience doubt that such a raft of infidels had to be stopped in their tracks?

Myers was joined in his special targeting of the Proposition 15 campaign by at least four of the other crusaders at the podium, including TVA chairman Aubrey Wagner, who identified Nader and his ilk as "demagogues," as "opportunists and publicity-seekers and self-appointed do-gooders." (He also called solar energy an "exotic" resource with a "two-percent efficiency," at best, compared with "32 percent" for a "coal-burning plant.") Paul Turner, vice-president of the Atomic Industrial Forum, contended that people like Nader "want to use nuclear power as a lever to bring down the establishment, to radically change it, to force their idea of a better lifestyle on the American people." While Nader "is one of the strongest, shrillest nuclear opponents," said Turner, "he has never spelled out the rationale for his opposition. And frankly I suspect he is motivated, at least partly, by his desire for a new kind of society. He recently gave an unusually candid interview to *Rolling Stone* magazine—this is a counterculture publication—in which he told what kind of society he wanted. The workers," said Turner, who must have been groping for his water at that point, "would run the manufacturing sector and the consumers would run the retail sector. Distributive and ownership justice—these were his words—would prevail over the wealth and technology we have even now to ensure that people wouldn't have to spend more than 20 hours a week in the economy. Like China, this society would get things done by marshaling massive numbers of people in a common cause." Turner concluded that "we all have a very serious interest in how these questions (of a nuclear moratorium) are addressed. Whether they are addressed in California, Oregon, Washington, or more near to Tennessee, they are going to have a significant and a serious impact on the future of this nation."

He should have been pleased, therefore, with a slide presentation by Joe La Grone, assistant manager of the San Francisco Operations Office of ERDA, wherein La Grone explained the action his office was taking—albeit illegally—to defeat the Proposition 15 initiative. La Grone said he wanted to "run through some of the highlights on the California nuclear power plant initiative in terms of where it is today, who is involved, what it looks like, what California looks like without nuclear power in terms of cost, the role that we have been playing, and some of the lessons that have been learned." Proponents of the initiative, said La Grone, included "Sierra Club people, People for Proof, Friends of the Earth, and so forth," while on "the anti-initiative side is a combination of the utility industry and organized labor, as well as nuclear vendors." (The latter were united in an organization called "Citizens for Jobs and Energy," co-chaired by former governor Pat Brown and William R. Robertson, Secretary-Treasurer of the Los Angeles Federation of Labor.)

Having described the "grass roots aspects" of the pro-initiative campaign, as opposed to the "traditional approach" of the anti-initiative side, La Grone outlined the ominous consequences of a nuclear moratorium in California, not the least being a five-fold increase in the cost of electricity by 1995. This "translates," he said, "into $7500 for a family of four." He added that "solar heating and cooling" might one day be a "competitive" option "in the extreme southern part of the state," and there "will be some help from geothermal, but by and large we can't expect much in the immediate future from these alternative energy sources." It follows, he said, that "for the near term we must rely on oil, coal, and nuclear power." As for ERDA's role, "we have been trying to convey facts, trying to convey the realities, explaining what options are available, what options will be coming on-stream at what time and at what cost." In particular, said La Grone, by way of clearing up "misinformation," ERDA had attempted to persuade the voters that "there really *is* a nuclear waste program." He was not pleased, however, with the results so far of the anti-initiative campaign, a recent poll having shown the election was "still a horserace." He told his audience that "somewhere, some of us are not doing our job for public awareness. Maybe it is happening in your state, too." He suggested that "the shot-in-the-arm approach is not going to work," as neither were "traditional methods of reaching the public through hearings, speeches, and press releases." What we need, he said in closing, is "a long-term educational program, and that involves you and all the people that are here. It is to your credit and your organization's credit that you have come."

La Grone understandably did not tell the whole story of ERDA's involvement in the Proposition 15 campaign. He didn't mention, for example, that while he stood there addressing the symposium a leaflet entitled "Shedding Light on Nuclear Energy" was being prepared by ERDA in Washington, D.C. specifically for distribution in California. Between February and April of 1976, according to a General Accounting Office investigation, almost 100,000 copies of the leaflet, which *Science and Government Report* (October 15, 1976) called "blatantly pro-nuclear," were shipped to "ERDA offices and major ERDA contractors" in California to be handed on to voters in the election. (Despite the GAO's contention that ERDA's brochure was "misleading" and tantamount to "propaganda," raising "questions about the agency's credibility," and despite clear proof that the leaflets were used for political purposes, the GAO did not call for legal sanctions against ERDA.) Likewise unmentioned by La Grone was the fact that his own office in San Francisco had mailed a letter to 500 civic, labor, and professional organizations in California, suggesting that they invite ERDA speakers to their meetings to discuss the impending election. Ironically, this violation of the law was uncovered and publicized by a Nader group in Washington, D.C.—but not in time to help prevent the narrow loss of the June 8 election to the anti-initiative side.

Nuclear power was thus assured a healthy future in California, at least for the time being, and the fate of the initiative there undoubtedly contributed to the subsequent loss of anti-nuclear elections in several other states that year, including Oregon, Montana, and Washington. This inspired an exultant Milton Levenson, director of EPRI's Nuclear Power Division, to proclaim in the *EPRI Journal* (January/February, 1977) that "the decisive results of seven state nuclear intitiatives during 1976 clearly indicate that any significant hazards of nuclear power are clearly outweighed (in the public mind) by its benefits." He added that "the continued increase in oil imports indicates that the basic objectives of Project Independence are not being met, and new nuclear plants coming on-line continue to be among the few restraints on the growth of oil imports." (Joe La Grone, meanwhile, in 1978—perhaps in recognition of service to his country—was promoted to manager of the San Francisco Operations Office, replacing Robert Thorne, a nuclear aficionado whom Schlesinger named Assistant Secretary of Energy Technology—with jurisdiction over solar programs—in the new Department of Energy.)

X
Utilities and Solar:
Hanging a Meter on the Sun

As suggested in previous chapters, the efforts of the utility industry to contain and hopefully to bury solar energy for the next couple of decades have not been confined to its mammoth competing programs in coal and nuclear power, nor to investments in the public relations and electoral spheres. (In 1977, America's 100 largest electric utilities spent over $60 million of their ratepayers' money on advertising and public relations campaigns—most of it, as discussed below, decidedly not in the interest of renewable energy resources.) A parallel tack has been to embrace solar energy with the two-fold purpose of restraining its development while shaping its applications to fit the plans and monopoly grids of the gas and electric utilities themselves. EPRI, for example, has fashioned a coordinated national "solar R&D program" for its 400-plus member utilities which affords the industry a critical degree of control over the future of solar energy at minimum expense to existing utility operations. "EPRI's modest funding of solar, geothermal, and fusion programs," writes Richard Balzhiser in the *ERPI Journal* (January/February, 1977), "enables the Institute to play a meaningful role in

ERDA's activities in these areas." Indeed, the Institute has "given ERDA constructive and necessary contributions as to the criteria new technologies *must* meet if they are to be accepted and provide maximum benefit to the consumers" (emphasis added).

One of EPRI's "contributions" to the solar industry is a study of *The Cost of Energy from Utility-Owned Solar Electric Systems*, co-authored in 1976 with ERDA and the Jet Propulsion Laboratory at Cal Tech. The document is subtitled "A Required Revenue Methodology for ERDA/ERPI Evaluations." It provides a logarithmic method for "the relative ranking of energy systems in a consistent manner," the purpose being "to relieve the problem of comparative evaluation of technology and plant concepts in the important developing area of solar energy." Adoption of the manual, at least by solar-interested utilities and research contractors funded by ERDA (DOE) or EPRI, is not optional: "All on-going and future studies by ERDA and EPRI solar energy system contractors will use the method; other energy system analysts are encouraged to do so as well." ERDA (DOE), meanwhile, "is planning to develop and release a companion model covering user-owned systems." This constitutes, in classic monopoly style, a foreordaining of the solar-electric systems to be installed and metered by America's utilities, and it chills the blood of people like David Marke, the Texas scientist whose modular on-site electric device may or may not meet the proffered "revenue" criteria.

EPRI's own involvement in solar-electric R&D has emphasized the power-tower, albeit grudgingly and chiefly to ensure that solar-powered electricity is amenable to centralized operation and control. Southern California Edison, one of EPRI's largest member-utilities, directs the consortium which will build and operate the government-funded experimental power-tower at Barstow, California, due for completion in 1981. EPRI itself has recently contracted with Boeing Aircraft and Black and Veatch Consulting Engineers, Inc. (known principally for its work on nuclear power plants) to refine a "gas-cooled" central-receiver concept which EPRI will test at Sandia National Laboratory. EPRI is further involved in a DOE contract with Stone and Webster Engineering, Inc. (another nuclear contractor) to determine the feasibility of centralized solar-electric power generation in the service areas of 11 southwestern utility companies. The Institute has also funded Stanford University to investigate the possibility of applying photovoltaic cells (clustered in "arrays" as large as half a square mile) to centralized utility functions.

All of which, again, is a bid by the utilities to influence the course of solar-electric R&D while otherwise pursuing business as usual. "Solar energy may have a place in electric utility applications," said an ERPI executive in 1975, announcing a series of Institute grants to Boeing, McDonnell Douglas, and other power-tower contractors, "but because of the nature of the sun's rays, that place will have to be complementary with conventional power plants, not at the exclusion of such plants (*EPRI News*, Fall, 1975)." The government has no quarrel with that, according to Michael Noland, a former Union Carbide executive now deputy director of the DOE-funded Solar Energy Research Institute (operated by Midwest Research Institute, a more recent employer of Noland.) In an interview with the *Energy Daily* (September 14, 1977), Noland says that even if solar power is contributing an unlikely ten percent of the nation's energy supply at the turn of the century, "it probably won't displace a single nuclear plant or coal-fired power plant." What solar will do, he adds, is "keep our standard of living better than if it did not exist."

This projection coincides handily with "facts and figures" disseminated by the electric utilities through anti-solar propaganda reaching millions of Americans every month. A clear example is the May, 1977 issue of *Aware* magazine (disguised mass-circulation PR voice for the utility industry), which carries an article by W. Donham Crawford, former president of Edison Electric Institute (EPRI's parent organization). After assuring his readers that the industry is doing its utmost on behalf of solar energy, Crawford swings about and declares that a "realistic estimate places solar electric generating capacity by the year 2000 at about one percent of the total installed capacity." (Precisely that figure was repeated in February, 1979 by a White

House task force commissioned to study the question, prompting an announcement over radio station WEEI in Boston, on February 27, that "solar energy may have potential as a power source, but it won't compete with coal or nuclear power in this century.")

Perhaps more saddening is the appearance of such "facts" in publications like the *Rural Electric Missourian*, distributed by the Association of Missouri Electric Cooperatives to its 260,000 rural customers. The July, 1977 issue of this magazine features a page of "Questions and Answers About Solar Power"—furnished, of course, by EPRI. "Sunlight is free," says the "fact-sheet," so "What makes its use so expensive?" Answer: "The fuel is free, but because it is diffused, large and expensive devices are required for capturing and converting it to electricity." Question: "Will solar plants ever replace nuclear and coal plants?" Answer: "No. Solar power plants may become competitive...for certain part-time applications," in which case "solar would be complementary but not competitive to coal and nuclear base-load power plants." Doubtless in consideration of its rural readers, EPRI was careful to include a "question" concerning "the potential of wind" as a source of electrical power: it is "being explored," says EPRI, "but cannot be counted on to contribute much to the energy picture for many decades. Wind is erratic and presents many technological problems." Finally, the standard EPRI/DOE forecast: "Assuming satisfactory technical progress and cost reductions, solar sources could produce 1% to 2% of our electricity by 2000."

It must not be forgotten, of course, that the millions of dollars in public money which the utilities and the government are spending on centralized solar-electric applications are directed as much at subsidizing aerospace corporations and keeping engineers employed as they are at solving energy problems immediately confronting the utilities themselves. In fact, the real solar threat to the utility industry are those applications which cannot be centralized, particularly the heating and cooling of buildings. It is a threat which perhaps bears heaviest on the gas monopolies (to be discussed in a moment), but nonetheless concerns the electric utilities as well. Since the 1950's, when they began hyping increased residential consumption of electricity through promotions like the "Gold Medallion" all-electric homes campaign, the electric utilities have drawn a steadily mounting percentage of their profits and growth from such consumption. They naturally hope to continue this pattern in the context of an expanded electrical grid—based on coal and nuclear power. Thus, in approaching the incipient national market for solar heating and cooling (SHAC) systems, America's electric utilities, with EPRI at the fore, have been obliged to wear the same two faces they wear in relation to other applications of solar power, namely to impede the growth of that market while appearing to promote it and gearing up to control it.

"Solar heating and cooling of buildings," writes Richard Balzhiser in the *EPRI Journal* (January/February, 1977), "could become competitive with conventional heating and cooling systems in the 1985-1990 time period and could penetrate 10% of the space conditioning market by 2000." EPRI's role, he says, is to help develop *"preferred SHAC systems* that not only conserve energy but also minimize the adverse impact on the utility system load factor by storing off-peak energy and using solar-assisted heat pumps." In order to identify what Balzhiser calls "preferred systems," EPRI has established a "SHAC demonstration program" in which the Institute and its member utilities sponsor the installation of commercial solar systems in selected model homes—usually amidst a fanfare of local publicity—and monitor the performance of those systems. EPRI has developed in conjunction with this a computer program that "permits determination of the preferred SHAC system for a specific utility service area." As of January, 1978, according to an EPRI survey, 124 local utility companies were sponsoring a total of 338 SHAC "demonstration" projects—funded largely by DOE under the terms of its "Memorandum of Understanding" with EPRI.

The effect if not the intent of most of these local "demonstrations" is precisely that of the ERDA/Honeywell Transportable Solar Lab: they portray solar power as a costly and exotic toy of the upper middle class, not as a viable meat-and-potatoes option for the average American working family. In Dallas, for example, the "SHAC demonstration" home constructed in 1976 by Dallas Power and Light Company was advertised for $90,000, of which $20,000 was entombed in a futuristic, bubble-domed solar collector system designed to provide no more than 30 percent of the home's annual heating requirements. That system, moreover, which EPRI doubtless "prefers," is based on a "solar-assisted electric heat-pump" concept locked into an otherwise quite conventional all-electric space conditioning apparatus.

A more extensive "demonstration" project was launched in 1976 by the New England Electric System, headquartered in Boston, with proportionately more damage inflicted on the evolution of a popular solar market. To design and manage the project, NEES hired a team from the Arthur D. Little Company headed by Dr. Peter E. Glaser (mentioned earlier in connection with the orbiting solar-electric satellite program). This team, culling from 5200 original applicants, selected 100 all-electric homes in the NEES service area to be retrofitted with solar hot-water systems provided by 18 different contractors and/or manufacturers at an average cost of $2000 per unit. (Participating households paid a token fee of $200 each, in return for which they could keep the solar systems when the project was completed.) Then, on July 13, 1977, having monitored the systems for one full year (spanning the vicious winter of 1976-1977), officials from NEES and Arthur D. Little summoned a Boston press conference to announce their preliminary findings.

On July 14, newspapers and wire services across the nation carried headlines similar to one in the *Wall Street Journal*: "Solar Heat Systems in Northeast Homes Prove Disappointing." *The New York Times* exclaimed: "Test Indicates Solar Heating Isn't Economical Yet," adding that a "preliminary look at the nation's first large-scale test of the efficacy of home solar water heating indicates that solar energy is not yet economically competitive with other forms of energy in the Northern states." Despite anticipated "savings of 50 percent or more," said the *Times*, "the average system efficiency was only 17 percent. The 15 best systems averaged a 37 percent energy saving, with the worst 15 averaging less than 5 percent." Solar advocate Christine Sullivan, Secretary of Consumer Affairs for the state of Massachusetts, was outraged by the negative publicity. She pointed out in her own press release that most of the defects in the solar hot water systems were minor and correctable, due primarily to inexperienced installers and clumsy supervision. Indeed, even the *Times* story ascribed the problems to "such things as air locks in water pipes, improper fittings, putting circulator pumps in backwards, faulty wiring...and leaks in pipes because of inadequate insulation." But, said Sullivan in an interview for this study, "the publicity damage was done, because the only things people read in a news story are the headline and the first few paragraphs—especially if the headline is negative."

Meanwhile, a NEES official who was also interviewed said his company would continue the solar demonstration for another 12 months, despite the fact, as he put it, that "even if the systems worked well, they would still take 20 years to pay back their costs." He added that of the 18 original solar manufacturers whose systems were being monitored, "four have been eliminated and one eliminated himself." (This was a reference to the president of a small Rhode Island solar firm whose system was ranked in the "worst 15" by Arthur D. Little. Shortly after the Boston press conference, whether because of it or not, he killed himself.) A few of the other NEES suppliers were local solar contracting firms, but the bulk of the systems were purchased either from national solar manufacturers or from solar subsidiaries of major corporations, including Daystar (Exxon), Sunworks (Enthone/ASARCO), Raypack, Grumman, Solaron PPG, and Revere Copper and Brass (ASARCO). The NEES official said finally that regardless of problems encountered in the demonstration program, the utility expects eventually to settle on a "preferred" SHAC system or systems which the company will lease to customers in its service area as it now leases conventional electric heating and hot water systems.

This would jibe with the government's approach to the matter, at least as sketched in a 1977 ERDA report which contends that the major barriers to a thriving solar market are "the high initial cost of solar heating and cooling (SHAC) systems and the disaggregated nature of the market" (meaning hundreds of small independent solar enterprises). It follows, says ERDA, that the best if not the only way to expedite the installation of SHAC systems is to turn the business over to the gas and electric monopolies, because they alone have the proper combination of capital resources, technical expertise, "quality control," and administrative machinery to tackle the job. Of the four different options which ERDA suggests for utility involvement in solar, the most extreme is that "public utilities be given monopoly franchises to provide SHAC systems." Here a utility would install and own the SHAC systems in a given "service area," billing its customers for such an amount each month as to amortize the cost of the equipment, pay for the sunlight consumed, and "return whatever rate of profit the (government) regulators would permit." A milder version of this proposal, which ERDA considers more likely to be accepted by the public and its regulatory panels, would deny the utilities a solar monopoly but allow them to install SHAC systems and charge for sunlight as an adjunct to their other "regulated" utility services.

California, it seems, has become a kind of proving ground for such ideas and for efforts by gas utilities in particular to monopolize the infant solar market there. In 1972, for example, with a grubstake of nearly half a million dollars in federal grants and $600,000 from its ratepayers, the Southern California Gas Company embarked on "Project S.A.G.E." (Solar Assisted Gas Energy), designed to "extend the life of natural gas supplies" through the installation of combined gas and SHAC systems in its customers' homes. The utility was "assisted" in this program by ERDA's Jet Propulsion Laboratory at Cal Tech, where an engineer named Alan Hirshberg and his associates devised a still more assertive scheme to install and monitor S.A.G.E. "demonstration" units in 315 residential and commercial establishments in the gas company's service area. To finance this proposal, gaily tagged "Operation Sunflower," SoCal Gas in early 1977 requested an $11-million customer rate increase from the California Public Utilities Commission, triggering a cannon-blast of opposition from consumer and environmental groups.

Among the more vocal of the plan's opponents was Peter Barnes, a San Francisco solar designer who confronted the PUC with a carefully reasoned, 38-page statement entitled "Utility Versus Consumer Ownership of On-Site Solar Energy Systems." The statement argues, first of all, that the PUC—to judge from its treatment of the hearings up to that time (April 14, 1977)—"is not really exploring *whether* the private utilities ought to be involved in solar energy development, but *how active* they have been" and should be in the future. Barnes scores the Commission for spending "considerable sums of money on consultants from the Jet Propulsion Laboratory—consultants who have a history of previous work with private utilities, and who unsuprisingly recommend large-scale utility involvement in solar energy development." Meanwhile, says Barnes, "the Commission spends no money...on consultants with a public interest background who might marshal facts and figures against utility involvement."

Barnes' own presentation methodically disintegrates every major point in the utility's case for "Operation Sunflower" (which would open the door to utility solar monopolies throughout California). Such a plan, says Barnes, would be neither less expensive nor more efficient than a vigorous state-supported program of consumer-owned solar installations purchased from small independent suppliers. Barnes points out that of the $11 million requested by the utility, only "$6.3 million is budgeted for actual installation costs; the remainder is for various kinds of in-house staff work." Assuming, says Barnes, that the entire $6.3 million were invested in "single-family domestic hot water installations at $2000 apiece, this would mean that for a total public subsidy of $11 million, the public would receive 3150 domestic hot water systems within five years." Barnes compares this with a "recently announced $4.6-million federal program to promote consumer-owned solar water heating systems in ten Eastern states" through direct grants of $400 to participating families. "Of the $4.6 million," says Barnes, "$600,000 is allocated for administrative costs. With the remaining $4 million, the federal government expects to stimulate the installation of

10,000 solar water heating systems within one or two years." (The federal program, while more efficient in the terms Barnes describes, has itself been harshly judged for concentrating the whole of its grant money in those northeastern states where several corporate solar subsidiaries, including Exxon's Daystar, have already gained a degree of hegemony over the retail solar market.)

Barnes argues further that the cost to ratepayers of utility-financed and installed SHAC systems would be "higher than the net price they'd pay as consumer-owners." Utility-free consumers "will almost certainly look around for the lowest price, probably by obtaining bids from a number of competing contractors." On the other hand, says Barnes, since "a utility's rate of return is calculated as a percentage of its rate base, it might well be in the utility's interest to install more costly systems. An imaginative utility might even be tempted to purchase its solar collectors from a wholly owned (and wholly unregulated) manufacturing subsidiary, much the way Pacific Telephone purchases its telephone equipment from Western Electric, thereby shifting ratepayers' dollars from limited-profit balance sheets to unlimited-profit balance sheets." In addition, says Barnes, the solar ratepayer's bill "is likely to include a chunk of the utility's general overhead—the large full-time crews, the meter-readers, the billers and clerks, the executives with their high salaries and expense accounts, the lobbyists, the political contributions, and a good deal of unnecessary advertising." To these must be added the noxious cost "of installing 'sun meters' on every utility-owned solar system, the cost of reading the sun meters, and the higher cost to taxpayers of supporting an expanded regulatory bureaucracy—costs that would be largely absent in consumer-owned systems purchased from small, competitive businesses."

To illustrate his point, Barnes contrasts the $2000 which SoCal Gas would charge to install a residential solar hot-water system with the $1500, at most, routinely charged by small firms. He concludes that the "long-term" difference in costs between utility-owned and consumer-owned SHAC systems "would amount to billions of dollars—billions of dollars that would be transferred from the pockets of consumers to the pockets of utility bondholders and shareholders." Aside from the larceny of such a "transfer," Barnes suggests that a fiscally superior means of obtaining capital for solar energy development would be to draw it from "decentralized" consumer resources— bank savings and credit arrangements, primarily—as opposed to the costlier "centralized" bond markets patronized by utility monopolies. "These are (consumer) dollars," says Barnes, "that might otherwise go for automobiles or recreational vehicles, new kitchens or swimming pools"—an enormous capital pool which "could be tapped for solar energy production, without coercion, by offering consumers one very reasonable incentive: the right to own the energy, and the dollar savings, that their solar investment produces."

Here Barnes raises the issue of consumer liberation from the nasty surprises of utility rate behavior. "The main attraction of solar energy," he says, "is the opportunity it offers to reduce future utility bills. If these future reductions are not in fact going to occur, or not be as large as consumers now expect, the incentive for consumers to invest in their own solar systems might well be eliminated." With consumer-owned systems, he adds, "future solar costs—like the cost of self-owned housing—can be calculated (and controlled) with a fair degree of certainty; these costs are a function of the cost of the installation, the cost of money, and the tax laws." With utility-owned SHAC systems, however, "future costs are considerably more unpredictable. The present thinking seems to be that the utilities' price for solar energy will be tied to the price of natural gas." If so, says Barnes, "as the average price of natural gas rises...any early price advantage of utility-owned solar energy would rapidly disappear." Indeed, "the track record of what happens to prices when energy corporations take over competing energy sources is not encouraging. During the 1960s, the major Amerian oil companies began acquiring large interests in coal and uranium. By the mid-1970s, coal and uranium prices have skyrocketed beyond all previous expectations."

Barnes adds that "California's experience with geothermal energy is also instructive. PG&E (Pacific Gas and Electric Company) operates one of the largest geothermal steam generating plants in the world at the Geysers in Sonoma County. But PG&E does not want geothermal steam to bring energy prices down. For nearly ten years, the nine northern California municipal power cities have been fighting to gain access to geothermal sites at the Geysers, and PG&E has blocked them all along the way," thus blocking any attempt to undercut the private utility's rates for geothermal power. Furthermore, says Barnes, "PG&E's lease contract with Union Oil—which owns many of the geothermal sites—provides that the price PG&E pays for steam shall rise in conjunction with the prices of competing fuels such as oil and uranium. In other words," he adds, "an oil company and a major utility are doing their best to make sure that geothermal power doesn't undersell their other sources of energy." He concludes that if "I were a member of the PUC or the Energy Commission, I would be extremely reluctant to sign over what may well be the last remaining competitive energy source—on-site solar energy—to monopolies whose proven inclination is to stifle inter-fuel price competition."

Barnes suggests a variety of methods by which utilities, in attempting to recoup their losses, might subvert the potential economies of consumer-owned SHAC systems. One approach would be a general rate increase for all customers. Another would be a "special 'stand-by charge' for solar users only," which the utilities would levy on the grounds that they "must maintain on stand-by enough back-up capacity to meet the needs of solar users during rainy or cloudy periods. Since the costs of maintaining this stand-by capacity cannot be met simply from the sale of electricity and gas during sunless periods, solar system owners would be made to pay an additional charge even when they are not using utility back-up energy." This reasoning, says Barnes, is particularly spurious for gas utilities, because "widespread use of solar energy would reduce—not increase—the required amount of natural gas storage capacity. Thus, utility overhead costs would be lower with solar than without it, and it would seem unreasonable to penalize solar users with a special stand-by charge."

Nonetheless, he says, utility companies are preparing such plans, and to prove it he quotes an "innocuous sounding, yet truly ominous, paragraph from PG&E's recent presentation to these hearings: 'The effect of increased solar utilization on the PG&E system will depend upon the nature of the solar installations and their resultant pattern of use. Solar systems may either increase or decrease existing load factors. Unless these changes can be correctly anticipated, *the potential cost savings may not be achieved and realized by customers.'*" (Barnes doesn't mention that the Public Service Company of Colorado in mid-1976 actually started billing solar customers for "stand-by charges." The outcry was so deafening that the company was forced temporarily to retreat. In 1977, however, according to the *Solar Energy Intelligence Report* of May 9 that year, an expected "prohibition against discriminatory rate scheduling... was dropped as unnecessary" from the final draft of President Carter's "National Energy Plan," again raising the "specter of punitive utility rates." It was also in 1977 that Consolidated Edison in New York City—when forced by a court-order to permit a small wind-system atop a Manhattan tenement to feed excess power into company lines—was allowed to charge that tenement higher rates for electricity consumed on windless days.)

Barnes' presentation included a section on the likely fate of "small, independent solar energy producers" in the face of large-scale utility involvement in the solar market. If, says Barnes, "the utility does not do the actual installations itself, but subcontracts the installations to independent contractors, small solar businesses could profitably, if somewhat parasitically, exist." The ability of those firms to market "self-owned solar systems," however, would almost certainly be hindered by the utilities' massive presence," not to mention the inherent "market advantages a utility company enjoys." A chief advantage is "automatic access to the pocketbooks of ratepayers," enabling the utility to beef up its solar earnings by simply commandeering "an across-the-board rate increase." Says Barnes: "No non-utility solar business has power like that." Another advantage is "monthly access by mail to every energy purchaser in the service area," plus "the image of dependability that comes from being large" and long-established, plus—perhaps most crucial of all—"the utilities' ownership and control of all the back-up systems to solar." A preview, says Barnes, "of what a utility-dominated solar marketplace might look like can be found by studying the telephone accessory industry since the *Carterphone* decision: while there are competitors selling customer-owned switch-boards and other hardware, the market is, and in all probability will always be, dominated by Ma Bell and its subsidiaries."

Besides overpowering and eventually displacing small-business competition, utility control of the solar market would hamstring technological innovation. "Monopolies generally," says Barnes, recalling testimony before the Senate small business committee in 1975, "and utilities in particular, are slow to innovate. They traditionally wait until a product and a market are established before plunging in, at which point their plunge is designed to knock out the early innovators and cash in on the long-term profits." Monopolies also have investments to pro-tect: "They are inherently unenthusiastic," says Barnes, "about pouring capital into new technologies that might devalue exist-ing investments before their full profit potential has been reaped. When monopolies do gain a foothold in new technolo-gies that compete with their not-yet-fully-rewarded investments, the tendency is to use that foothold to slow down the introduc-tion of the new technology rather than to speed it up. Thus, AT&T, a major participant in COMSAT, has not exactly done its utmost to promote the use of satellites for long-distance communications. Similarly, the oil companies have done very little (prior to 1978) to develop the Western coal leases our government so innocently granted them not long ago."

It isn't that Barnes perceives no role at all for utilities in the growth of a solar market in California. He includes a suggestion for them in the list of recommendations with which he closes his statement to the PUC: "Encourage private utilities to develop renewable energy sources that are more suited to centralized ownership, e.g. solar-steam generation of electricity and pro-duction of methane from agricultural wastes." But like a burn-ing star at the top of his list is Barnes' insistence that the PUC "prohibit private electric and gas utilities, or any subsidiary thereof, from owning on-site solar energy systems." A compan-ion statute would forbid the utilities from "levying a 'special stand-by charge,' or any other charge which would discriminate against solar users as opposed to ratepayers generally." Barnes further suggests that "the Commissions and the legislature do everything possible to promote investment in consumer-owned solar systems." Topping the list here is "tax parity for solar consumer-producers," aimed at equalizing the tax burden on private solar energy "producers" relative to the utilities which "receive extremely favorable tax treatment in the United States." (PG&E, according to Barnes, "paid federal income taxes of just 2.9 percent of its pre-tax profits of $242 million in 1975.")

A strong first step toward solar tax-parity, says Barnes, would be to eliminate the state sales tax on "solar energy collecting equipment"—since, he says, other "energy producers are not subject to the sales tax on their equipment purchases." Correlative provisions would "exempt consumer-owned solar systems from property taxes" and "increase the existing 10 percent state income tax credit (for SHAC systems) to 50 percent," extending the credit "to passive as well as active systems." Barnes would further apply "the six percent sales tax to final sales of natural gas," a measure he believes would (1) inspire gas conservation, (2) buttress the incentive to "go solar," (3) generate state revenues for possible use as low-interest loans to solar consumers. "Very appropriately," says Barnes, "the tax (on natural gas) would make users of a diminishing resource finance the gradual introduction of that resource's logical replacement." With regard to low-interest loans for the purchase of SHAC systems, Barnes emphasizes that the "goal in both public and private loan programs would be to make the monthly finance charge for energy conservation and solar energy less than the monthly savings on utility bills—from the very first month." Finally, Barnes suggests "mandatory" thermal efficiency standards for building construction and design, "consumer education" (including solar vocational training programs), and consumer protection" (including "sun-rights" legislation, performance standards for solar equipment, and government licensing of solar technicians).

* * * *

It is possible that the California legislature heard Barnes' message, at least with one ear, because five months after he presented his statement to the PUC, Governor Jerry Brown signed legislation providing a range of new incentives for the purchase of consumer-owned SHAC systems. The chief incentive is a lifting of the state income tax credit to 55 percent of a solar investment, up to a maximum of $3000 during the next four years. This credit applies to "retrofits" of existing buildings as well as to new construction, though the California Energy Commission expects the heaviest use of the credit in connection with new home sales in "solar-powered" housing developments. Attached to the law are mandatory insulation standards and performance criteria for solar equipment, which must be approved by state inspectors. Brown has also appointed a citizens' advisory board called the SolarCal Council, charged with the formulation of a comprehensive solar development program for California. (Among the members of this board is radical activist Tom Hayden, whose own national political organization—the Campaign for Economic Democracy—has advanced the notion of a solar transition as one of the major planks in its platform.) A significant factor in the passage of these measures was the strong support of younger state officials, including Governor Brown, who has ordered solar installations in a number of state-owned buildings. Such enthusiasm within the government, coupled with projected increases in the cost of natural gas in California, is expected to help the tax-incentive boost solar installations to 170,000 units by 1981, according to the state Energy Commission.

This legislation is admirable. It will doubtless be used as a precedent by other states and even the U.S. Congress in pursuit of much-needed solar policies and incentives. Considered by itself, however, the legislation in California does little to avoid or retard an eventual monopolization of the state's solar heating and cooling market by gas and electric utilities in alliance with major "energy" corporations. Meanwhile, those utilities have shown no indication of withdrawing their bid for such a monopoly, toward which evidently they will continue to be "assisted" by the federal government. SoCal Gas, for example, is still trying to promote "Operation Sunflower," and Alan Hirshberg, the aforementioned engineer at ERDA's Jet Propulsion Laboratory, believes that the major impediment to such programs is not "consumer activists" so much as the reluctance of the utilities to push aggressively into the solar market. According to an item in *Politicks and Other Human Interests* (October 25,1977), Hirshberg is convinced that "the idea of utility ownership of solar systems," which he helped conceive, should and will eventually come to pass. (An intriguing postscript: Hirshberg left the Jet Propulsion Laboratory in 1978 to join the Washington, D.C. engineering and consulting firm of Booz-Allen and Hamilton, which lists as clients a number of the nation's largest utilities, as well as the federal government. In 1977, the firm received $2.6 million in contracts from DOE alone. In March, 1978, Hirshberg played a major role at a government-funded Washington conference on "The Solar Energy/Utility Interface.")

XI

CORPORATIONS AND SOLAR:
BIG FISH EATING LITTLE FISH

It will be recalled at this point that one of six key items in the corporate solar development strategy outlined earlier is to "mesh the burgeoning solar market into the larger corporate market by absorbing small successful firms (or their ideas), emphasizing mass production, and placing distribution under the control of utility companies." Having now considered in some detail the other five items in that strategy—from centralized control of capital-intensive R&D investments to deterrence of the public from viewing solar energy as an immediate popular option—it is time to look more closely at what in a sense was the original impetus for this study: the displacement of small enterprises from the solar market and the appropriation of their ideas by member institutions of the corporate elite.

There is, of course, nothing new in the phenomenon of small-firm displacement by larger firms. It is simply a form of growth, and growth is inherent in the capitalist system. That is why, indeed, the small family farm, the corner grocery and the local bank, the family-owned cotton gin and virtually every other small private enterprise—once the life-blood of U.S. commerce and manufacture—have vanished from the American landscape. In the late 19th and early 20th centuries, write Barnet and Müller in *Global Reach,* the "same techniques now being used to develop the global market—modern transportation, accounting, marketing, etc.—were successfully employed to create (an) integrated U.S. market, and in the process local pockets of resistance to the march of the great corporations were eliminated." The advent of the global business enterprise has merely accelerated this engine of concentration. Hence, "during the course of the last generation the top 500 corporations have dramatically increased their position of dominance in the American economy. In 1955, 44.5 percent of all American working in manufacturing and mining worked for the top 500 corporations; by 1970, the figure had risen to 70 percent. In the same period, the top 500 increased their share of all manufacturing and mining assets in the country from 40 to 70 percent."

The authors add that the "rhythm of accelerated concentration is sustained by the perpetual process of merger," a process which by the late 1960's had reached "an exponential rate." In 1965, for example, "1496 domestic (national) firms disappeared through merger, the highest number in the history of the United States up to that time." A principal instrument of merger and concentration is "cross-subsidization—i.e., the use of power and resources developed in one 'profit center' to start or to expand another. The cross-subsidization strategy is used within the United States by electronics firms to conquer the bread market or by banks to become buyers and leasers of aircraft."

Carried far enough, as it has been in the United States, this inexorable drive toward growth through concentration will lead to what Barnet and Müller (borrowing from Robert Averitt) call a "dual economy": the " 'center' economy, comprising a few hundred firms, controls over 60 percent of the productive and financial resources of the country," while the " 'periphery' economy is made up of thousands of smaller firms dependent on the giants for their survival." Moreover, write the authors—and here is perhaps the crucial point with regard to concentration in the solar industry—"the effect of government spending (especially in such areas as defense and energy) is to accelerate the process of concentration," because the "benefits of government spending go overwhelmingly to the 'center' industries." The subsequent influx of government cash, along with orders for goods and services, only enhances the ability of the "center" corporations to absorb smaller enterprises in the "periphery" economy.

Certain aspects of this phenomenon in relation to solar energy development have been examined in previous chapters, pratically through the tribulations of solar entrepreneurs like Jerry Plunkett, David Marke, and J.H. Anderson. But there are other important aspects—pertaining specifially to "cross-subsidization and "centrist" concentration—which are best understood through the following case-histories of small solar firms and inventors who have felt the weight and occasionally the spite of large corporations attempting to "cross-subsidize" a path for themselves into the U.S. solar market. Furthermore, in tandem with the establishment of a corporate Solar Energy Industries Association, there is the critical and apparently foreboding Wall Street connection.

"A growing number of giant U.S. companies are becoming linked to the emerging solar industry," says a writer for *Solar Engineering* (December, 1976), official publication of the Solar Energy Industries Association (SEIA). The article refers to several different "paths of entry" for companies adding solar to their product lines and sales activities. Some corporations, says the author, "such as Olin Brass, Revere, and Kennecott, supply basic materials to (solar) manufacturers," chiefly in the form of metals, glass, and surface coatings for solar collectors and system components. Other suppliers include Alcoa, Dow-Corning, Owens-Illinois, and ASARCO (the later to be discussed in a moment).

Firms may also enter the solar market through the "manufacture of products which facilitate the use of their tooling in the manufacture of solar hardware. This occurred with Champion Home Builders, the largest producer of mobile homes in the U.S.," which now manufactures a complete prefabricated "solar house" and Champion Solar Furnace distributed via the company's 3000 dealerships nationwide. "Chamberlain Manufacturing," says the author, "with its expertise in handling aluminum and glass in storm windows and doors" has obtained a license for "the Winston compound parabolic collector, which will be manufactured next year for use in producing industrial process heat." Grumman Aerospace has likewise retooled for solar, as have (to name but a few) Carrier, Lennox, PPG, Hamilton-Standard, General Electric, Westinghouse, and General Motors, whose Harrison Radiator Division in 1977 converted a large plant from radiators to solar hot water systems.

An allied "path of entry" here is the addition of solar equipment and services to the inventories of national retail chain operations. Both Sears and Montgomery Ward, according to *Solar Engineering*, are actively involved in the testing of solar domestic hot water systems. Due to the competitive nature of the market, however, neither will announce yet when the products will be introduced." (A more recent development involving Sears is treated in the following chapter.) Meanwhile, a number of corporations "with product lines traditionally directed toward the heating and refrigeration industry now see a potential for their products in the solar industry with some modifications. Among those firms from the controls industry are Honeywell, Robertshaw, and Johnson Controls," as well as the Bell and Gossett Division of ITT, which "recently produced a catalogue directed toward the solar market, showing the use of such products as switch meters, manual valves, and flow control valves in solar systems." The West German firm of Grundfos, "world's largest manufacturer of pumps, has marketed directly to the solar industry, and their pumps are evidenced in several system packages. Grundfos recently opened its first U.S. manufacturing plant in Clovis, California."

By and large, notes the author, solar-interested corporations have found "the most difficult path of entry" to be the innovation and market introduction of new solar products. It is partly for this reason that such corporations have welcomed the growth since 1974 of federal spending on solar R&D (recalling the *Business Week* article quoted earlier with regard to "building a new business at tax-payer expense"). The author points out that Lennox Industries "quietly captured the solar equipment for over $600,000 worth of installations for the most recent HUD demonstration grants," helping the firm "come to the forefront in developing and marketing a complete solar energy system." And "PPG industries (a glass manufacturer) took an early lead in the industry with its Baseline Solar Collector, which is being used in several government demonstration projects." Another firm with success "in getting its products specified on government demonstration projects" is Revere Copper and Brass, "now one of the largest producers of modular collectors and domestic hot water systems." Grumman Aerospace, too, while "establishing dealerships for its Sunstream solar systems," has also "received government contracts for wind energy research and has introduced a small wind generator."

These examples along reveal the muscle of the corporate state as it lumbers into a nascent solar market. Its advantage over smaller "competitors" is obvious: tremendous capital resources and control over raw materials, established factories and dealer networks, brand-name identification, unlimited advertising budgets, and high-level access to government R&D funds. That should be quite sufficient to assure the giants whatever degree of hegemony over the new solar market they desire. (Indeed, most of those very corporate attributes are cited by government energy planners as justification for transferring the nation's solar R&D program to the large corporations, the presumption being that only they, in concert with the utilities, have the wherewithal to execute the "commercial development" of solar energy in the United States.)

And yet, says *Solar Engineering*, on top of the "paths of entry" described above, it is "the purchase of existing companies (which) has put many energy companies into the solar industry." Among the reasons, of course, is simple capitalist economics: a corporation wishing to diversify (especially into a field rife with uncertainty and unfamiliar technology) will often find it cheaper and less demanding to purchase a struggling but promising smaller firm, along with its patents, technologies, and even personnel, than to build a new corporate division from scratch.

Such a purchase may also bring to the corporations a freshness of product which its own market research experts would be hard pressed to emulate. (That sort of motive was apparently at work in an attempt by Dow-Corning in 1977 to purchase an interest in Steve Kenin's New Mexico Solar Room Company, which uses a Dow-Corning product in the fabrication of its polyurethane "skin.") And finally, the acquisition of a smaller concern is the surest and probably the least expensive way to avoid competition from the enterprise in the future.

It would take some digging to find a more classic example of the tightening corporate grip on America's energy future, including the bantam solar industry, than that provided by ASARCO, formerly the American Smelting and Refining Company. With revenues last year in excess of a billion dollars, ASARCO isn't a corporation so much as a conglomerate of corporations, the bulk of which are engaged in one form or another of mining and metals processing. ASARCO placed an ad in the September 27, 1976, issue of *Barron's Weekly* which proclaimed: "Electric Power from the Sun—A Reality with the Help of Silver." The copy alluded to the government's massive power-tower program due to consume up to half a million square meters in heliostats whose reflective properties derive from a coating of silver—"the world's largest producer" of which is ASARCO. The firm ranks fifth in the world production of copper as well, explaining its majority holdings in Revere Copper and Brass, which abides in turn among the nation's top ten manufacturers of solar collector systems (with Revere's featuring copper absorber plates, naturally).

There's more: ASARCO is a major producer of cadmium, whose unique metallic behavior is under development by Shell/SES, Texas Instruments, and other firms in pursuit of commercial photovoltaic cell production. Meanwhile, in late 1975, through its Enthone copper subsidiary, the conglomerate purchased Sunworks, Inc., a formerly modest California solar operation which uses a copper oxide collector surface developed by Enthone. Twelve months later, when HUD announced its "Second Cycle" of solar demonstration grants, Sunworks received a total of $219,000 in contracts, while Jim Piper's solar firm got $7000. Piper didn't bother to apply for "Third Cycle" funding, announced in July, 1977, but Sunworks/ASARCO did, and collected more than $1 million for its trouble. In 1978, the firm won contracts totaling more than $1.5 million in HUD's "Cycles 4 and 4-A," making Sunworks/ASARCO one of the top five solar manufacturers in the United States.

Another leading "energy" corporation which has bought a piece of America's solar future created by someone else is Mobil Oil. In 1972, according to an unpublished book by Gerald Schaflander called *The Fight for the Sun,* Dr. Bruce Chalmers of Harvard University was working in a cramped,underequipped laboratory on a potentially important breakthrough in photovoltaic cell technology. The value of Chalmer's long-researched process, now known as "Edge-defined Film Growth" (EFG), was the promise it held for drastically reducing the cost of manufacturing solar-electric cells—the dream technology in solar R&D since they were first deployed as power systems on American space craft in the 1960's. Most solar cells are "sliced" by hand from batches of specially processed silicon, a technique so expensive that commercial cell arrays—used primarily to fuel electric devices such as warning lights in remote locations—cost up to $15,000 per peak kilowatt (the output of the cells at high noon on a cloudless day). This compares with $750 to $1000 per peak KW from conventional electric plants fired by fossil fuels at current (1979) prices.

The key to reducing the unacceptable cost of silicon cells, hopefully to $500 per peak KW or less, is to automate the production process, eliminating the arduous, time-consuming work of "slicing" silicon wafers by hand. Dr. Chalmers' EFG technique purported to accomplish that by "growing" silicon cells in a continuous "ribbon," one-eighth inch in width, fed onto a spool from the furnace where the silicon is processed. But all Chalmers had in 1972 was a theory and a tiny grant from NSF. "He was using a small laboratory," writes Schaflander, "and did not even have a spectroscope, which is vital for measuring silicon" and similar substances. It was then that "Chalmers made a connection with Tyco Laboratories, a Waltham, Massachusetts corporation with $30 million a year in sales. In return for the use of their well-equipped laboratory, he agreed that their vice-president for research and development, Dr. A. I. Mlavsky, would become a Junior Investigator under Chalmers, working on an NSF grant for EFG." Thereafter, according to Schaflander, "Mlavsky and Chalmers' rapidly advanced Chalmers' original work into a batch-processing technique...producing up to seven-inch-long ribbons of solar cells." Problems remained, however, and the research team in mid-1974 enlisted the aid of scientists at ERDA's Jet Propulsion Laboratory in Pasadena, California.

Meanwhile, says Schaflander, Tyco had been seeking additional funds, and suddenly, "in a bold entrepreneurial action," the corporate lab "sold Chalmers' invention, financed and (ostensibly) controlled by NSF, developed by Chalmers and Tyco, to Mobil Oil Company—for 20 percent of the stock in a new $30-million solar corporation. Mobil controls 80 percent of the stock, and Dr. 'Eddy' Mlavsky, is executive vice-president." Amidst the shuffle, Chalmers dropped so profoundly out of sight that Mlavsky, in a six-page interview with *Solar Age* (April, 1976) concerning the origin and development of EFG, doesn't once mention the Harvard scientist's name. (This may be due in part to Chalmers' continued association with Schaflander himself, who in a long-standing feud with Mobil Oil accused the firm in a letter printed in *The New York Times* [January 31, 1977] of deliberately trying to suppress the development of solar cell technology.) Mlavsky does, in the interview with *Solar Age*, talk a good deal about himself, pointing out that it was "on my initiative, I would say," that Tyco Laboratory "adopted as its mission...the concept of converting technology into businesses rather than merely into reports."

And why, asks *Solar Age*, did the lab sell its EFG technology to Mobil? Tyco recognized in 1974, says Mlavsky, that EFG would require a further R&D investment which the lab could not afford: "Thirty million dollars was the number we felt was necessary." In a "poker game," he adds, "if you sit down to play, and you draw two or three aces in your first few cards at stud, you're going to lose to a pair of deuces if you don't have enough money to follow the next bet. To recognize this, early, is a very important piece of corporate wisdom. So," he says, "we began to look for a partner. Now what should be the characteristics of such a partner? It had to be someone who could put a great deal of money into development, where the outcome was still speculative...That partner for a number of reasons turned out to be Mobil Oil Corporation." Did the man say "speculative?" Six months after the interview, *Solar Age* noted in its "Business Beat" section that stock in Mobil-Tyco, listed on the New York Stock Exchange, "has been showing price gains on the heels of a most favorable earnings report. The stock rose to 15⅛ after the company reported sales for the first quarter of its current fiscal year of $37 million—up from $11.1 million for the same period in fiscal 1976—and a net of $1.3 million, or 45 cents a share, up from last fiscal year's $668,000, or 22 cents a share."

A fate not dissimilar to Dr. Chalmers' was visited in 1975 on another potential pioneer in solar-electric cell technology. The scientist is Stanford R. Ovshinsky, who owns a small firm in Troy, Michigan called Energy Conversion Devices, Inc. Unlike Dr. Chalmers, Ovshinsky is devoid of academic certification, not even a bachelor's degree, and that evidently is a noticeable factor in his troubles. "Part of Ovshinsky's problem," writes *Washington Post* columnist Nicholas von Hoffman, "is that he has the ability but not the credentials, and it has been hard for school men to accept the thought that someone who has never been to college has made an important contribution to science." In early 1977, Ovshinsky announced that he had conceived a new method for treating "sheets" of silicon cells that would lower their cost from an average of $15 per square foot to 50 cents, thus producing electricity for as little as 0.2 cents per kilowatt hour, compared with the current three to five cents. "The implications are enormous," said Ovshinsky at a press conference, according to *Business Week* (July 18, 1977). The magazine added that while there was as yet no prototype, "Ovshinsky says that a device could be on the market in three years. 'It's a matter of investment, not invention,' he insists."

But the scientist is unpedigreed. Ten years earlier, Ovshinsky had made an equally dramatic announcement concerning a breakthrough in amorphous semiconductor materials which he claimed would "revolutionize" the electronics industry. Due to his brazen style and absence of credentials, Ovshinsky's claims were discounted by the industrial scientific community, and he was unable to raise the money to develop his concept. Now, says *Business Week*, the Burroughs Corporation is "manufacturing amorphous semiconductors for some of its computers, and some engineers believe that the unusual properties of the devices could make them extremely valuable in certain applications." Indeed, says David Adler, an MIT professor quoted in *The Wall Street Journal,* "On every single point of controversy it is now clear that Ovshinsky has been correct from the very beginning, and it is about time that the scientific community acknowledges this explicitly."

Such acknowledgment has not been forthcoming, and Ovshinsky continues to get a cold shoulder in his attempts to raise money for his silicon cell technology. One party who turned him down was John Coburn, technology development manager for Exxon Enterprises, Inc. Coburn told *Business Week* that "I don't see how you can produce electricity for less than 20 times the cost that Ovshinsky is talking about." (It may or may not be relevant that Exxon in 1973 acquired a photovoltaic subsidiary called Solar Power Corporation.) Meanwhile, says *Business Week,* Ovshinsky's firm "remains as it always has—a company struggling to stay afloat." In 1976, "Ovshinsky was forced to sell 50 percent of the rights to any solar devices to United Nuclear Corporation, a uranium producer." All of which brought von Hoffman to add in his column on Ovshinsky: "His real difficulty is that he is a throwback to the one-man-inventor-genius-manufacturer of a century ago. The corporate world has put a stop to free-lance da Vincis like Ovshinsky. If the man had gone to college and become a proper, organizationally subservient, modern genius, he would have been sold by his deans and professors into the bondage of a corporate laboratory, where the fruits of his invention would have been signed over to United Gain, Inc."

Another "free-lance da Vinci" who has had experience with Exxon Enterprises recently is a Texas inventor named Dan Schneider. Since 1973, when he gave up a flourishing medical practice to pursue his invention full-time, Schneider has been fighting to enter the market with a unique and cost-effective "lift-translator" device which will generate electricity both from wind and from flowing water. There has never been a question as to the workability of Schneider's machine, only resistance by the federal government and private investors to the "unconventional" nature of the technology involved. This, in turn, has left Schneider vulnerable to attempts by other companies to appropriate his machine for their own ends, whether those be to bury the machine or to stamp it with their logos and promote it in the marketplace.

Instead of using propellers or turbines to "catch" the latent energy in wind and flowing water, thereby driving a generator, Schneider's "lift-translator" employs the "lift" effect of a series of "vanes" shaped like airplane wings and attached to a chain-drive apparatus resembling an elongated vertical ferris wheel. The machine will "start up" in winds as low as four miles per hour, will tolerate winds as high as 80 miles per hour without "feathering" or disintegrating, and will yield a steady flow of power at fluctuating intermediate ranges. The wind-machine can be applied on any scale from one to 1000 kilowatts or more, while the water-driven version will operate effectively in "heads" (or waterfall heights) as low as two feet and as high as 200.

With regard to cost, Schneider has designed a wind-machine for a site near Gatesville, Texas due to produce up to 100 KW of peak power in 20-mph winds for an installed total of $53,100. A second machine for a customer in Arizona will generate 40 KW for $25,900—provided Schneider can stay in business long enough to fill the order. (By way of contrast, DOE has recently spent $1.3 million for a 200-KW wind-machine from McDonnell Douglas and $1.5 million for a 40-KW machine from Lockheed and Westinghouse.)

Not only has Schneider himself built and tested prototypes of his machines, including a pair of hydroelectric devices for the government of South Korea, he has submitted his design to outside laboratories for further verification of performance. Among those labs are the St. Anthony Falls Hydraulic Laboratory at the University of Minnesota and, more recently, the Davis Hydraulic Laboratory at the University of California.

Both labs reported an efficiency of 90 percent or better for Schneider's low-head hydro-machine in a variety of simulated applications ranging from dams and sluiceways to coastal, river, and ocean currents. Yet Schneider has been consistently rebuffed in his attempts to win development contracts from the wind-electric and low-head hydro divisions of NSF, ERDA, and DOE.

These rejections are the more disturbing in view of the care which Schneider has taken to propose to the government an actual market application for each device needing government support. "I don't like to ask for government money," said Schneider in an interview for this study, "until I'm sure of the practical role of a machine in the marketplace." Hence, in submitting an application to the low-head hydro program at DOE in mid-1978, Schneider worked out a cost-sharing arrangement with a small utility in Rochdale, Massachusetts which intended to use the Schneider machine as part of its commercial operation. In December of that year, Schneider received a letter of rejection from DOE asserting that: (1) Schneider's machine did not "represent a technological improvement" over conventional generating turbines; (2) Schneider's machine was no less expensive than conventional turbines; (3) Schneider's utility partner stood to earn $22,995 per year from the machine and thus should bear a heavier proportion of design and installation costs.

Schneider was appalled. In an angry letter to Stanley I. Weiss, the DOE official who rejected his proposal, he pointed out that: (1) his machine was an improvement over conventional hydroelectric technology in five specific and significant ways, all documented in a government-funded evaluation by the University of California at Davis; (2) the cost of the Rochdale installation, because it was to be a "one-up" *prototype* built by hand, could not be usefully compared with the cost of conventional generating machines "that have been mass-produced for 50 years!" Schneider might also have noted, with regard to the Rochdale utility's potential "revenues" from the machine, that such considerations have not troubled the government in its massive "solar demonstration" awards to Coca Cola bottling plants, Xerox office towers, and Budweiser breweries—each of which will pocket hundreds of thousands of dollars in fuel-cost reductions.

Schneider is aware that he is playing with loaded dice in a corporate-government craps game. So long as he was willing to restrict his R&D efforts to very small hydroelectric machines— below a head of 15 feet—the government was willing to lend him a modicum of support (in the form of a contract to evaluate his design at UC/Davis). But the Rochdale installation would have called for a machine in the "15 to 50 foot" range of heads. "My device has no competition at all at heads below 15 feet," said Schneider," because turbines just cannot be cost-effective in that range. But as soon as I move out of the niche that DOE has made for me and start competing with existing technology—at the 15 to 50 foot head range—they don't want me to get into that, because there I'm competing with Allis Chalmers and the other big operators that have been in there for 50 years or more." (He might have mentioned Lockheed, in addition to Allis Chalmers, since Stanley I. Weiss, the man at DOE who sent him the nasty letter, was until September, 1978 vice-president for advanced programs and development at Lockheed's Space Systems Division in Sunnyvale, California.)

At any rate, it is this kind of treatment by the federal government which has left Dan Schneider in the uncomfortable position of a lone inventor with orders piling up for his machines which he cannot fill for want of development capital. (He is presently operating out of a backyard factory at his home in Irving, Texas, where his neighbors have mounted a campaign to force him to move.) Ordinarily, he says, given his backlog of orders, he would be able to attract conventional investors to his enterprise, but those he has approached are afraid to risk their money on a technology for which there is no established market.

"I've *had* to seek DOE support," said Schneider, who is genuinely pained at having to turn to the government, "because as long as I don't have a favorable position with the Department of Energy, I can't find risk capital on the outside." Except, that is, from Exxon Enterprises and Pittsburgh-Desmoine Steel, Inc., both of whom have offered to capitalize the further development of Schneider's technology in return for controlling interest in his company: Exxon wanted 90 percent, Pittsburgh-Desmoine 51 percent. Indeed, when Schneider refused the offer from Pittsburg-Desmoine, the company went to court in a futile attempt to circumvent his patents. "I'll tell you what I told Exxon," said Schneider at the close of his interview: "I am not going to give my invention away."

While Schneider tramps the Texas plains in search of investors and a place to work, a solar company in Massachusetts which Exxon Enterprises did support and then acquired is expanding "on schedule." In fact, the narrative to follow is not so much a case of corporate vamping on a brilliant unruly inventor as a study in the kind of thinking that underlies "success" in the field of solar energy—a field which many had assumed five years ago would not be subject to the traditional corporate definition of "success." It is also a story which further and garishly typifies the hidden benefits reaped by corporations from federal R&D programs—filtered through corporate research labs and favored universities—pursuant to a "high-tech" energy future for the United States.

The focus of the story is a pleasant and apparently innovative research engineer named Gary Nelson, senior vice-president of Daystar, Inc. in Burlington, Massachusetts. Nelson was employed until 1972 by the Itek Corporation, a Massachusetts firm whose specialty at the time was thermodynamics and optical systems research in support of the federal aerospace program. It was in 1972, of course, that the war in Vietnam was staggering to a close and the government was shrinking its aerospace budget. "When the program started to decline," said Nelson in an interview for this study, "we had a closetful of technology in the lab at Itek with no application for it." So he and a colleague, in sniffing about for possible applications, hit on solar collector design as a means of using their expertise both in optics and in thermodynamics. By late 1973, they had built in the company lab a prototype collector system with a unique corrugated "heat-trap" feature which gave the device an unusually high efficiency. During this period, said Nelson (which encompassed the Arab oil boycott in 1973), "we realized that there was an energy problem and maybe a market for solar collectors." They offered the system to Itek management as a commercial option, but Itek declined, allowing them instead to pursue commercialization on their own.

"We formed a company in 1974," said Nelson, who was president of the firm, "and started looking around for financing." They were funded initially by a "venture capital group" and then "negotiated a contract with Exxon Enterprises to show them we had some technology that might be of interest. Exxon was favorably impressed and asked if we would be interested in equity financing." Daystar accepted. "Regardless of what people think," said Nelson, "it takes capital to get going in the solar market. It's sort of a paradox. Everything is being directed toward small business, but the capital investment to do this kind of job and meet the economics is huge." According to Nelson, "the relationship with Exxon developed further, and they did more investing until finally at this time we are a wholly owned subsidiary." He said the corporation had "turned out to be good investors to work with. They understand the needs and the technical problems and they understand the methods by which you have to operate to develop a business. If you take a look at the major solar companies today, you find that all of them now are affiliated with large corporations because that's the only place today that you can get capital."

Daystar's earliest customers in 1975—and still its principal source of contracts—were utilities and government agencies, including the Department of Defense. "Let me put it to you this way," said Nelson: "the major force, the major determinant in the solar industry today is and will be for a good while the federal government, directly or indirectly." Asked about the million-plus dollars spent on Daystar systems in the "Third Cycle" of HUD demonstration contracts, Nelson denied that his company was "directly contracted by the government. We have more than 90 dealers around the country," he said, "and they identify construction projects that meet the objectives of HUD demonstration programs. They propose a certain application using our equipment and our systems. Each of them works out a separate grant arrangement with HUD."

Most of Daystar's authorized dealers, said Nelson, are

"east of the Mississippi, but we're not limiting ourselves to that in the future." The company has "a very, very strict dealership program" with "stringent requirements," including an "extensive technical and marketing training program" at the Daystar plant in Burlington. Nelson's dealers "install and service only Daystar solar systems," he said, "except in areas that we aren't involved in, such as swimming pool heaters." Guided by Exxon, Daystar has projected an orderly schedule of company growth and expansion which Nelson appears to be satisfied with. "We're a mass producer now," he said, "although we're not operating at top capacity yet." Indeed, the Exxon-financed Daystar factory is turning out collectors at only "25 percent of potential capacity," resulting in a "significant backlog" of orders for Daystar systems. Still, said Nelson, "we're operating at our programmed expansion rate. We gauge ourselves by design to respond to the market requirements in a reasonable growth manner rather than to go gangbusters."

Of course. And will the company build other plants eventually? "Yes," said Nelson, but he doesn't know when, because the solar industry is "new and none of the normal marketing or projection techniques apply because there's no historical data to go by. Everyone's trying to sort out how big the market is and where it is and when it will go." Does Exxon plan to make solar energy a major part of its corporate future? "I think," said Nelson, "that Exxon is facing the same situation that everyone is facing, including the United States government. The solar industry *can* grow and *should* grow—" Nelson stopped, then added: "I'm not speaking for Exxon, so I guess you shouldn't quote me on that. *If* solar becomes an acceptable industry—and many things must happen before that point—then it can become an industry. I think the Exxon people recognize that fossil fuels are limited, so they have to be an energy company, not an oil company. I think everyone in the business of conventional energy has recognized they must be diversified. Solar," he concluded, "is promising and has its place, but it's not the total solution."

* * * * *

Exxon, then, has acquired control of a superior solar product developed by aerospace engineers in a government-funded corporate lab, has used its connections in Washington to subsidize the growth of its acquisition to a point where it already shares domination of the solar industry with ten or twelve other major manufacturers, and has written an "expansion" program to suit the needs of its other corporate interests. Among those interests, and one which illustrates what Nelson must have meant by "energy company," is a "laser-fusion" research project at Rochester University (read Rockefeller) apparently aimed at putting Exxon into the business of producing plutonium for nuclear breeder reactors. It is not a small program, and a brief discussion thereof is not irrelevant to the subject of corporate acquisitions in the solar field.

Since 1972, according to *Energy Futures*, (the ERDA-funded industrial survey above cited), Exxon Research and Engineering has pursued a new laser technology in partnership with General Electric, Northeast Utilities, the State of New York, and, of course, ERDA (DOE). In 1976, says *Energy Futures*, "Rochester University entered into a $15.1-million agreement with ERDA (DOE) to build a laser six to ten times larger than the present one. The private participants, including the university, will together contribute $24 million to the project. Exxon will also provide manpower for operating the facility...in Rochester."

The stated intent of this research, which involves the smashing of microscopic pellets with an ultra-potent laser beam, is to advance the technology of fusion reactors (touted by proponents as a "safe" application of nuclear power). But in fact, says *Energy Futures*, "Exxon generally believes that fusion will have no impact on the nation's energy supply for at least 20 years." It follows that the company is more intrigued with the laser's potential for "rejuvenating spent fuel rods from fission reactors, which contain mostly U-238 (plutonium)"—a technology Exxon believes "can be proven within five years." If so, says Dan Grafstein, Exxon's laser project manager, "the earliest application for laser fusion will be to produce plutonium to fuel breeder reactors." And it happens, according to *Energy Futures*, that a nuclear subsidiary of Exxon in Washington state "already capable of producing plutonium fuels" has "been inhibited by technical problems which the Rochester project is intended to solve."

Exxon, then, while adding a key producer of solar energy systems to its petroleum, coal, and synfuel operations, is also investing millions of dollars in precisely those nuclear technologies whose centralized character and thirst for capital have thrown them into headlong competition with the needs and promise of small-scale solar energy. This brings a new shaft of light to bear on the corporate acquisition of solar energy firms, and the light becomes a glare in view of the fact that perhaps the majority of solar-interested corporations are also interested in nuclear power. ASARCO, for example, with its Sunworks and Revere solar holdings, mines and markets uranium as well. United Nuclear Corporation owns 50 percent of Stanford Ovshinsky's solar inventions. Mobil, General Electric, Honeywell, Bendix, Westinghouse, Rockwell International, General Atomic (Gulf), and a host of other "energy" corporations with significant nuclear commitments have likewise diversified into the fledgling solar industry. The unsavory implications of this trend became especially visible in late 1976, when Sunburst Solar Energy, Inc., of Menlo Park, California, was acquired by EDS Nuclear Corporation. The latter firm, according to *Solar Age* (December, 1976), "provides technical and management assistance to companies building nuclear plants and earns over 90 percent of its revenues from electric utilities or their contractors. EDS, says Sunburst president Larry Newton, is negotiating the company's first solar-utility consultant contract, and the tie-in should work out well. In the long run, Newton believes, the utilities must get involved, as 'they are sitting right in the middle of the whole problem.' "

Solar Age contends that the Sunburst acquisition by EDS caused stock in the nuclear company to "hit a new high" on the New York Stock Exchange. Whether true or not, that the magazine should say so points up the existence of a recent and very significant Wall Street connection: the more obvious it becomes to America's investment community that *Fortune 500* corporations with nuclear and fossil interests are assuming control of the solar market, both sanctifying and defining its future, the more willing is that community to finance the process. "In the past three years," writes Helene Kessler for Pacific News Service (July 4, 1977), "some 100 firms on the public stock market have introduced solar technologies, either as their principal products or as new divisions of older companies. Today, two to three companies with solar products register on the stock market each month." This "sudden growth," says Kessler, "has even created a new niche in the securities profession—the solar investment adviser."

One of these is Anthony Adler, whose "Solar Investments Associates in Mamaroneck, New York advises some 200 clients." Adler himself, writes Kessler, "by following his own advice on solar investing has averaged a 250 percent return." Richard Livingstone, a New Hampshire analyst "who started writing his 'Solar Energy Stock Letter' a year and a half ago, today has more than 1000 subscribers." Kessler adds that among "the latest investment firms to enter the solar market is the Hirsch Organization in Old Tappan, New Jersey, the third largest investment firm in the country," which "advises clients on solar investments through its publication *Smart Money*." This firm's "solar expert, Ronald Rotstein, is currently working on *The Handbook of Solar and Alternative Technologies*, a guide to more than 40 alternative energy technologies and the 100 solar companies now on the stock market."

It is doubtless not entirely by chance that many of the solar-interested corporations recognized as such on the New York and American stock exchanges likewise constitute the core membership of the Solar Energy Industries Association (SEIA). Founded in 1974 by its current president, Sheldon H. Butt, Director of Market Research and Planning for Olin Brass, the organization is headquartered in Washington, D.C., where it lobbies alongside most other national industry associations, including those representing coal, petroleum, and nuclear power. Among the "Corporate Members" of SEIA are Alcoa, the Aluminum Association, Daystar (Exxon), Dow Chemical, Edison Electric Institute, Enersol (Southern Union), General Electric, Grumman, Honeywell, ITT, Lennox, McDonnell Douglass, Owens-Illinois, PPG, Revere Copper (ASARCO), Rockwell International, Solar Investors Associates, Solar Utilities Company, Southern California Edison, Sunburst (EDS Nuclear), Sunworks (ASARCO), and Northrup (ARCO).

Equally suggestive are the hundreds of less auspicious solar entrepreneurs who have not kicked in the $100 to $8000 in annual dues required for membership in SEIA. They include most of the small manufacturers profiled in this study, from Jerry Plunkett, Dave Marke, Steve Baer, and Steve Kenin to J.H. Anderson, Peter Hunt, Dan Schneider and Jim Piper. It was Piper who observed in an interview that "I've always considered SEIA to be an organization of materials suppliers. They could care less about solar energy." While that is perhaps a trifle harsh, there is no reason whatever to doubt the complicity of the organization in a "hard-path" energy strategy among whose aims is continued corporate traffic in "advanced technology" and profitable raw materials such as copper, glass, aluminum, and brass. The number one "purpose and objective" of SEIA, according to its membership directory, is to "accelerate national and international development and marketing of solar energy conversion systems." A modest success in this regard was the organization's role in a decision by ERDA in 1977 to spend $4 million "stimulating" sales of solar hot water systems in 10 northeastern states (hunting ground of Daystar, Sunworks, General Electric, PPG, Revere, etc.). "At a March 28 press conference," says *Solar Engineering* (published by SEIA), "Dr. Henry Marvin [of] ERDA's Solar Division acknowledged the suggestion for such an initiative from SEIA last summer."

Another of the self-assigned functions of SEIA—this one meriting careful scrutiny by advocates of a diverse and independent small-scale solar market—is the introduction of a uniform body of solar equipment specifications and performance standards, presumably in the interest of "consumer protection." The organization has established a "Solar Energy Research and Educational Fund" (SEREF), which in 1977 was contracted by

FEA to help develop a "rating and evaluation" system for commercial solar equipment. George Szego, chairman of SEREF, told *Solar Engineering* that he expects SEREF's "recommendations on solar equipment to be looked upon as a combination of Underwriters Laboratories for safety and *Good Housekeeping* for value and recognized performance standards." The secretary-treasurer of SEREF, John C. Bowen, is also chairman of the "Standards Committee" of SEIA, which is heavily dominated by representatives of corporations and associations engaged in the marketing of basic manufacturing materials. Olin Brass is represented, as are Revere Copper and Brass, Ametek, Inc., Reynolds Metals, ITT (Fluid Handling Division), and the Copper Development Association.

There is a disquieting potential for abuse here, and it is heightened by the likelihood that SEREF's "evaluation" procedures and performance criteria for solar equipment will be adopted by such organizations as the National Bureau of Standards and the American National Standards Institute, both of whom are principal sources of building code data and construction specifications for thousands of American municipalities. In addition, the National Bureau of Standards has launched a program of "Round Robin Solar Collector Tests" to be conducted by 31 government and corporate laboratories, including Arthur D. Little, General Electric, Honeywell, Lennox Industries, Lockheed, Martin Marietta, NASA, and Union Carbide. ERDA (DOE) has commenced a similar program in which seven private laboratories will "evaluate" the solar systems of "each manufacturer who has sold 10,000 square feet or more of solar collectors," according to ERDA's "Request for Proposals."

It requires little imagination to see the possible effect of these "evaluation programs" on small independent solar manufacturers. Suppose, for example, that Steve Kenin's "solar room," which by NASA criteria wouldn't be considered a "solar collector" at all, is not included in a SEREF/NBS "Approved Systems Checklist" mass-distributed to potential consumers of solar equipment, or, for that matter, to consumer credit establishments making loans for such equipment. Suppose alternatively that a neighborhood construction cooperative seeking to build its own solar systems confronts a local building code which specifies that anything less than a "high-efficiency" collector system made of aluminum or copper will require a 100-percent conventional "back-up" heating system.

Lest such hypotheses appear unlikely, consider the experience of a Texas inventor and earth materials specialist named Harold Skaggs, who in 1975 was restrained by local building code officials from constructing a house he had designed with dried adobe blocks instead of conventional brick or concrete. It took the designer seven costly weeks to obtain a waiver of the code, despite the fact that no structural problems were involved, merely regulations and the will of the code officials. A similar incident occurred in Florida in 1976 when tinkerer Jim Wurth was refused a city permit to erect in his backyard a pair of small electric wind-turbines which he had purchased from a farmer in South Dakota. This dispute dragged on for nearly a year. It was finally resolved in favor of Wurth when his neighbors submitted to the city council a strongly worded petition on his behalf. (Wurth had also been stymied by officials of the local electric utility, who were so unhappy with his turbines that eventually they purchased and removed them). In subsequent remarks to the Associated Press (June 26,1977), Wurth said that "someone trying to do something innovative to conserve energy shouldn't get the hassle I got."

It is in such a manner that building codes, performance standards, and professional licensing laws can be and have been used to advance the interests of a powerful commercial group against the interests of less powerful competitors—hence one's discomfort at the thought that SEIA will exert a major influence on codes and standards for the solar industry. While the bias of the organization is clear enough from its membership roster, it has been further clarified by a peculiar series of interactions with U.S. Representative Mike McCormack. In 1975, for example, SEIA elected McCormack its "Solar Man of the Year," presenting him a "solar-powered" wrist watch in appreciation of his attempts to push through Congress that year a bill to fund the installation of 4000 solar heating and cooling "demonstration" units. McCormack brandished his solar timepiece before a television audience in Washington state as he predicted, according to *Solar Energy Washington Letter* (April 28, 1975), that "there will be no significant total energy savings from solar by 1990 and only a 'small amount' by 2000. Meanwhile, he said, nuclear power will advance to 30 percent of U.S. electric energy supplies by 1985 and 60 percent by the year 2000. McCormack pointed out, however, that the one percent to one and a half percent of total U.S. energy consumption provided by about seven million solar homes in the next decade would be a vital contribution to our energy picture."

The link between McCormack and SEIA grew still more bizarre in July, 1977, when McCormack distributed to the entire membership of the U.S. Congress copies of a book entitled *Soft vs Hard Energy Paths*, published by one Charles Yulish, a New York management and communications consultant widely known for his activities in support of what Amory Lovins calls "the electronuclear industry." In fact, Yulish had assembled his book from ten solicited essays attacking Lovins for his "soft-path" energy article in *Foreign Affairs*, and one of those ten essays was written by Sheldon Butt, president of SEIA. McCormack declared in a letter accompanying the book that "the issues raised in this discussion are fundamental to America's future. They are of greater importance than is generally realized. I urge you to take this report with you on your next trip (or wherever you have a chance to read) and take the time to read it."

Butt's contribution to the volume, preceded and followed by contributions chiefly from apologists for nuclear power, opens with an expression of "profound shock" at Lovin's argument in *Foreign Affairs*. "The basic justification for our technological society," writes Butt, "lies in the exceptionally broad opportunity it has provided to the citizens for individual expression and the exercise of individual initiative." Butt is persuaded that such "opportunities" could not exist apart from the material base of a "hard" technology supporting an economy of abundance. It is "implicit" in Lovins' proposal for a "soft-path" energy future, says Butt, "that the population must sacrifice a great deal in terms of material well-being." He challenges Lovins' contention that an enlightened majority of Americans would favor a "soft" energy path: "the central issue of the (1976) election," he says, "was economic," and "the winners were those who convinced the voters that their policies were the ones best suited to accelerating economic growth." Lovins, he adds, in calling for a "new society founded upon the concept of 'elegant frugality,'" is calling for "a society of peasants and craftsmen" which America will reject.

Thereafter, in a gesture most striking for the president of a solar energy association, Butt attempts systematically to discredit the potential of solar energy—at least as applied before 2025. He blasts a suggestion by Lovins that autos could be fueled on corn alcohol instead of gasoline. He cites a table of government figures to dispute Lovins' assertion that 100-percent solar heating in the U.S. would under any forseeable circumstances be cheaper than electrical heating by 2000. He froths at the gills over Lovins' proposal to substitute "passive" solar systems, which function without hardware, for the "active" systems produced by Grumman, Daystar, and Olin Brass. This idea, says Butt, "carries us one step further forward into the realm of fantasy." (In 1974, at a New York "Solar Collector Workshop" sponsored by NSF, Representative McCormack warned against an energy policy based on "solar fantasies.") Butt closes by saying that "we in the Solar Energy Industries Association are dedicated to actively pursuing the commercialization of solar utilization within the framework of our present society based upon individual election and individual initiative." Due, however, to the capital demands of other deserving energy technologies, solar will provide, by 1990, only about "4 percent of our total budget" for heating and cooling buildings (i.e., less than .05 percent of America's gross energy requirements) "and ultimately"—by 2025—"about 60 percent of these needs or 12 percent of our total budget."

Not only, therefore, have Wall Street corporations thoroughly "penetrated" the U.S. solar market through intracorporate diversification ("cross-subsidization"), extensive government subsidy, and the purchase of smaller firms, they have organized a solar industries association clearly devoted to building a solar market that will be compatible with the larger aims and "hard-path" energy goals of the corporate elite in general. While it would be impossible to calculate the market strength thus accumulated by the large corporations involved, there is mounting evidence to suggest that in less than five years they have concentrated sufficient control over the solar industry to squeeze out smaller competitors and effectively prevent the entry of others.

Three years ago, for example, in Austin, Texas, an entrepreneur named Frank Hutchinson and his partner Jean Spence decided to combine their technical and administrative skills in the formation of a solar collector manufacturing company. Upon investigating the national solar market, however, they found, according to Hutchinson, that "the major corporations were already sitting on the sidelines. They already had the manufacturing (of solar hardware) pretty much sewed up." Spence and Hutchinson opted instead to become what they call "solar design-builders," installing the equipment of major manufacturers in structures of their own design. Having attended a two-week training course at the Exxon/Daystar factory in Massachusetts, they are now part of Daystar's network of 90-plus dealers. "The independents aren't going to make it," said Hutchinson. "Two years ago there were 17 solar-only companies in Austin. Today there are five, and three of them attached to national corporations."

This corresponds with the findings of a recent DOE survey of solar manufacturing activity—the sixth such survey conducted by the agency since 1974—which reveals that fewer and fewer solar manufacturers are producing an ever higher percentage of solar energy systems. Of 186 leading firms surveyed, 81 produced or imported over 90 percent of the 1.9 million square feet of medium-temperature collectors manufactured in the first six months of 1977. A mere 18 of those firms produced 46 percent of the medium-temperature collectors, while 15 produced *more than 90 percent* of the 3.2 million square feet of low-temperature collectors (the type generally specified in HUD demonstration contracts). Most of the other key indicators in the DOE survey point unmistakably toward a classic concentration of corporate power in the young solar industry. In its first five semi-annual surveys, for example, DOE found sharp increases in the number of new firms producing enough square footage of collectors to be included in the survey—from 39 companies in 1974 to 102 in 1975 to 177 in December, 1976. The latest survey, however, recorded a net increase of only nine firms.

Speculating on this phenomenon in the June, 1978 issue of *Solar Age* magazine, Allan Frank pointed out that "one characteristic revealed by the survey is the apparent increasing stability of the industry. What causes this apparent stability remains uncertain. Is it a settling of the industry or a reluctance of prospective participants to enter until (federal) tax credits are approved and the market picks up again? The evidence indicates that it may be the former. Few (new companies) entered the field in 1977—and few dropped out." Meanwhile, a ranking utility executive in Oakland, California, quoted in a 1978 publication called *Jobs from the Sun*, spoke bluntly about the chances of survival for small solar entrepreneurs. Addressing proponents of public intervention on the side of small business, the executive said: "You *know* these small businesses can't survive. They don't have the payroll, the inventory, the capital expansion budget. You *know* that five years from now there won't be more than 20 solar businesses in California."

The foregoing treatment of these developments is not to suggest that the involvement of major capitalists seeking a profit from solar investments is inherently a negative thing, at least in terms of the immediate future of solar and allied renewable fuel resources. It has been observed more than once in this study that the rapid deployment of solar technologies will in fact require an enormous capital investment, and, to paraphrase Jim Piper, it is better that a capitalist profit from energy conservation than from energy waste and damage to the environment. But there is a profound distinction to be observed between capital invested at a moderate rate of profit on behalf of a general social good and capital invested with an eye toward personal and corporate profit above all other conceivable "returns" on that investment. It is the latter perception of profit which sustains the corporate class—indeed has given rise to its birth and evolution as the dominant force in American society—and it is the venerable Wall Street investment apparatus which constitutes the purest expression of the corporate understanding of profit. One is brought hence to conclude that Wall Street participation in commercial solar energy development, on the scale and in the fashion described above, is the clearest possible signal that the corporate-government program for controlling the destiny of solar energy in the United States has met with considerable initial success. It is likewise a signal that small-scale solar energy in the hands primarily of small producers, aiming at decentralized, democratic, and maximally efficient applications, has been strangled in the crib.

XII

THE PEOPLE AND SOLAR: TOWARD
COMMUNITY SELF-DETERMINATION

The wizened reader may now be asking: what else is new? After Vietnam and Watergate, after the mayhem in Santiago and the horror at Three Mile Island—how could anyone be surprised that the government of the United States has sanctioned a global energy program directed at the further concentration of wealth and power in the hands of the corporate elite? What reasonable observer could have expected from the government a solar energy development initiative that would break the nation's 40-year stride down the road of constant economic growth and "progress" through technology? By what twist of logic and political acumen could the small-scale solar energy community have thought that its hopes for a "husbanding approach toward resources, land, and eco-systems" might be taken seriously in the chambers of government and the corporate boardrooms of the nation? "Today," write Barnet and Müller in *Global Reach*, "most populist attacks against bigness, banks, and interlocking directorates have a faintly quixotic air since the U.S. economy has been hurtling for so many years in the opposite direction."

191

There is no disputing this, and yet there has been awakened in the land a sense of crisis regarding human survival itself which overrides cynicism and fuels an admittedly naive faith that those in charge of the nation's destiny will recover their vision before it is too late. "Oil and natural gas," to repeat the quote from Denis Hayes, "are our principal means of bridging today with tomorrow, and we are burning our bridges." Amory Lovins writes that we "stand at a crossroads: without decisive action our options will slip away." Further stalling on the deployment of "diverse soft technologies pushes them so far into the future that there is no longer a credible fossil-fuel bridge to them: they must be well underway before the worst part of the oil-and-gas decline." Thus it is that "enterprises like nuclear power are not only unnecessary but a positive encumbrance for they prevent us, through logistical competition and cultural incompatibility, from pursuing the tasks of a soft path at a high enough priority to make them work together properly." The "conditions" for building a "soft-path" energy future, writes Lovins, "will not be repeated. Some people think we can use oil and gas to bridge to a coal and fission economy, then use that later, if we wish, to bridge to similarly costly technologies in the hazy future. But what if the bridge we are on now is the last one?"

These are among the survival realities which make the "success" of the hard-path corporate energy program discussed heretofore a potentially disastrous failure of the government to function in the interest of the mass of its people—to function, indeed, as "a government of, by, and for the people" themselves. The dimensions of this failure can best be appreciated by glancing now at what is being lost to us in the absence of a systematic, accelerated national campaign to reconstruct the energy grid of the United States in favor of decentralized renewable technologies and resources.

The driving force of a soft-path energy arrangement is a physical principle which might be called "the logic of the ecosystem." It is in the nature of this "logic" that it will differ in its major characteristics from ecosystem to ecosystem, from that of Duluth or Sacramento, say, to that of Houston or Santa Fe. But whatever the nature of the specific climatic, geologic, and terrestrial conditions producing a particular ecosystem, it is the holistic combination of those conditions which determines the "logic" of that ecosystem, and it is that "logic" which in turn dictates the nature of the energy technologies and resources to be exploited in an ecosystem without inflicting damage upon it.

Closely related to the "logic of the ecosystem" is Lovins' principle of "matching" a particular energy resource as exactly as possible to the desired "end-use" of that resource: a dentist's drill requires electricity or compressed air while the drying of clothes requires a cotton cord or length of wire strung up in sunlight. The coupled observation of these two principles will lead *ipso facto* in the direction of "appropriate technology," and from the deployment of appropriate technology will spring *ipso facto* the following returns to the human community and its environment:

1. minimum waste of resources;

2. minimum damage to the ecosystem;

3. minimum capital and operating expenses for energy production, with savings especially dramatic in the transport of energy from point of production to point of end-use (hence the illogic of centralized electric plants located scores and even hundreds of miles from the light bulb needing the power);

4. maximum energy self-sufficiency at each progressive scale of use, from family to factory to neighborhood, city, region, state, and nation;

5. maximum capital and financial self-sufficiency at the community level (assuming the availability of local "capital pools" as discussed by Peter Barnes in Chapter X);

6. maximum stimulation of locally owned and operated small business enterprises;

7. maximum stimulation of employment opportunities, due to the labor-intensive character of appropriate energy technologies (flat-plate solar collector manufacturing and installation produce more than five times as many job/years per dollar invested as nuclear plant construction and maintenance, according to *Solar Age*, August, 1978);

8. escalation in community "standards of living" due to

surplus capital and resources derived from the economical primary allocation of those resources;

9. positive changes in the scale and design of human settlements and patterns of daily life, tending toward smaller urban concentrations separated by farmland and wilderness, simplified transportation systems, work/home living arrangements, home and community food-production systems, and a generally stronger emphasis on individual participation in community affairs;

10. reduced alienation between individuals and the fundaments of nature in their ecosystems, hence a new reverence for nature and for living, life-producing things;

11. reductions in crime against persons and property, due both to factors mentioned above and to greater equality of "income" and "lifestyle," these changes due, at least in part, to the tendency inherent in small economic institutions to yield but modest personal wealth;

12. improved public health, due to a cleaner environment, more nutritious food (locally grown with a minimum of processing), and reduced anxiety and stress;

13. higher degrees of individual and group participation in the *political* process, due to higher degrees of individual and group control over the *economic* process determining the quality of people's lives;

14. enhanced prospects for world peace.

These desirable conditions of life, abstract and utopian as they may appear in their compressed format above, have lately caught the eye and pricked the imagination of millions of Americans, a massive proportion of whom expressed themselves clearly during the national Sun Day celebration in the spring of 1978. Indeed, local and national organizers of that celebration have called its success "a message to Washington from the people of this country saying we want solar energy now!" It is common knowledge that the broad public interest in solar development is producing a heavier volume of letters and phone-calls to members of Congress than almost any other single issue.

Furthermore, a growing network of small but vocal solar advocacy organizations has been unleashing ever-sharper critiques at the federal solar program, some of them demanding the resignation of the entire senior executive stratum of the Department of Energy.

Two different questions emerge from all this. One: is there any evidence to suggest that Washington has heard the "message" of Sun Day, 1978—any indication in the form of new programs or policy directives that the federal government intends to make small-scale solar energy development a major item on the national agenda? Two: if not, what possible recourse have the people themselves to fill the void left by the failure of their government to act on their behalf? The second question is not one simply of whether the people of the United States—indeed of the global biosphere—shall enjoy the elegance of a soft-path energy future that enriches their lives while helping to secure their long-term survival. It is a question bearing on issues raised in the Declaration of Independence, issues of liberty and self-determination: how much longer will the people of this country endure the tyranny of the corporate elite? How much longer will they accept the dimestore gew-gaws, flashing lights, and plastic biscuits of the *Fortune 500* in exchange for control over the force, direction, and quality of their existence as human beings? How much longer will they fail to remember that their ancestors once would have risen in revolt against the sorts of encroachment on their means of self-reliance—their farms, their cabinet shops, and the sovereignty of their communities—which now are matters of routine expropriation by the corporate state?

Concerning the first question: by far the closest thing to a recent federal gesture toward a democratized solar energy program was a White House "Domestic Policy Review" of the existing government program announced by President Carter in his May 3, 1978 "Sun Day" speech at Golden, Colorado. Referring to solar energy as "a cornerstone of this nation's energy policy," Carter said he was ordering the DPR as a prelude to stepping up federal solar development efforts. He also claimed that he was adding an unexpected $100 million to the federal solar budget for fiscal 1979. Both of these pledges were more or less kept, but only in a manner strictly unthreatening to the larger aims of the hard-path corporate energy strategy. The supplemental $100 million, for example, according to *Energy Daily* (May 17, 1978), was filched "without too much pain" from insignificant coal and nuclear projects which the Department of Energy had terminated well before Carter's Sun Day speech. More than half the money came from a scuttled plan to construct additional nuclear waste facilities, and $45 million was shifted from a program to demonstrate "clean-burning" coal as a boiler fuel. Once transferred to the solar program, moreover, the bulk of the $100 million was assigned to solar-electric technologies, including wind, photovoltaics, and low-head hydro, with an humble $5 million each for "appropriate technology," "dispersed energy systems," and "passive solar heating and cooling."

The "Domestic Policy Review" was completed in December, 1978 after 12 public hearings in as many cities and several months of apparently tortuous deliberation by the 30-odd federal agencies participating in the process. An early draft of the DPR was blasted by Amory Lovins as a "largely vacuous and uncreative exercise in ducking the big issues," which for Lovins included the question of "market parity" for solar energy and the inappropriateness both of solar-electric power towers and of "silly" big wind machines. Ken Bossong of the Washington-based Citizens Energy Project wrote in *Ways and Means* magazine (November-December, 1978) that while "the DPR may yet prove to be a turning point, it is no longer apparent that the turn will be in the right direction." Bossong points out that "all proposals (at the 12 public hearings) for communty-oriented solar projects have been dropped" from the DPR, along with "shifts of at least some solar R&D funds to small business." Instead, writes Bossong, "the options focus heavily on utilization of private investor-owned utilities for commercializing solar energy; a strong federal role in setting national solar policy; and solar commercialization in suburban areas at the expense of cities and rural communities."

The final draft of the DPR, though bereft of genuine programmatic responses to these objections, at least admits for the first time ever in the history of the federal program that it is possible to achieve a 20-percent solar contribution to the U.S. energy supply by the year 2000. There is a snag, however: it will cost the government between $49 billion and $74 billion *in the next five years* to attain that figure, according to the analysts who wrote the draft. A second option proffered by the DPR would have the government "spend" $2.5 billion from 1980-1985—counting $1.5 billion in the form of consumer tax credits—for a solar input of 10 to 15 percent by 2000. It is the second option which Secretary Schlesinger has recommended to President Carter, meaning in effect that there will be no increase in federal solar spending above the level of $500 million per year inscribed in DOE's FY 1980 budget long before completion of the DPR. This comes to the same five percent of the federal energy budget that solar has received since 1978, versus 44 percent in 1980 for nuclear power (including the weapons program) and 27 percent for petroleum (including the billion-barrel "strategic reserve" which caught fire in Louisiana late in 1978—among its other difficulties).

More important than total dollar-figures, of course, is the question of who will decide how the solar funds are spent on what technologies and programs in whose apparent interest—the question which has occupied most of this book. The answer is the same as it was in 1971: the corporations and utilities of America will decide how the solar funds are spent through their emissaries at DOE and other federal bureaus. Not only has the corporate line-up at DOE not changed since Carter's Sun Day speech, it was bolstered in April, 1978 by the hiring of a "super-salesman" named Jackson Gouraud to spearhead Schlesinger's much-prized "technology commercialization" program. "During his career in the marketing departments of Remington Rand, Pfizer, Seagrams, and Liggett and Meyers," says *Energy Daily* (June 23, 1978) of Gouraud, who also runs his own consulting firm, "Gouraud figures he has handled somewhere between 180-200 different products, and provided advice to perhaps 100 companies."

Now, says *Energy Daily*, Gouraud "provides advice to Dale Myers, DOE's undersecretary in charge of technology development. Energy Secretary James Schlesinger brought Dale Myers in from North American Rockwell (Rockwell International) to make the technology side of the agency behave like an industrial marketing organization. Myers, in turn, has brought in Gouraud as his deputy secretary for commercialization. If Myers is the general manager of DOE," adds *Energy Daily*, "then Gouraud is Myer's marketing manager." Below Gouraud in the chain of command are some 20 hand-picked "resource managers," each of whom is being trained by Gouraud to supervise the "market penetration" of a new technology deemed by Schlesinger to be ready for "commercialization." Among these "marketable" technologies are "enhanced oil recovery," "coal gasification," "passive solar," "wood," "solar hot water," and "cogeneration" (of electricity from industrial process steam).

The "central constraint on a technology like cogeneration, Gouraud believes," according to *Energy Daily*, "is the number of entities that must be brought together to make it happen— the right large industrial facility, the equipment supplier, the local utility, perhaps various state offices and federal agencies." Gouraud contends that this "is really a sales engineering job at quite a high level. These (resource managers) are going to have to deal with the companies at the board level." *Energy Daily* points out that "Gouraud plans a similar sort of thrust in solar hot water systems, save that he will tap the four regional solar energy research institutes (SERI) for his sales engineering staff. With solar he is aiming to plug into existing channels of distribution, like the heating oil dealers in New England, for example, with whom he met earlier this week. 'I've told the head of the eastern regional SERI that I look to him to get the fuel oil dealers organized,' says Gouraud. In solar," says *Energy Daily*, "Gouraud expects the field sales force will eventually number over 100 people."

What *Energy Daily* has failed to report is that Schlesinger's top-down "technology commercialization" strategy, with Dale Myers and Gouraud at the controls, has generated chaos in a number of programs at DOE and driven some of its ablest staff to resign. One of them is a bitter Don Elmer, the geothermal official quoted in Chapter VI with regard to MITRE and other

"Beltway Bandits" in Washington, D.C. Until October, 1978, Elmer had been working on a geothermal implementation program involving close liaison between DOE and the governments of five western states with geothermal resources. "Because I am a populist and an old-line Jeffersonian democrat," said Elmer in an interview for this study (February, 1979), "I was always trying to push the responsibility and the management control down to the lowest levels, even past the state governments, past counties and cities down to the individual citizen if I could—but at a minimum to the city-county level."

After more than a year of saucer-balancing negotiations concerning the use of the western region's geothermal steam reserves, Elmer was poised to sign a multi-state, multi-county agreement for joint state-federal development of those reserves "with a depth of information sharing," said Elmer, "going all the way down to personnel, budgets, programmatic design, and contract approval." One week before the scheduled inking of that agreement, said Elmer, "our [DOE] division was suddenly reorganized and all my policy and liaison functions had disappeared." This blow was dealt by geothermal director Rudy Black under orders from Jackson Gouraud, whom Elmer called "one of the world's few perfect idiots. His [Gouraud's] claim to fame is designing special bottles for Schmirnof liquors that have never sold. He did a Viking bottle for vodka of which there are tens of thousands on dealers' back shelves—and now he's DOE's commercialization manager." Elmer recounted a departmental briefing of labor leaders on "technology commercialization" that was orchestrated by Gouraud: "Not once," said Elmer, "did he mention the job implications of any of the technologies, and he was only saved from his stupidity by the fact that most of the labor leaders were even dumber than he was—they didn't ask."

Elmer argues vigorously that his approach to "participative planning" for energy resource development "is the only workable approach. If you involve the broadest possible spectrum of people in the planning process, the solution is *their* solution, and you don't have to sell it to them afterwards." Furthermore, "by reaching past the bureaucracy to the people in their communities, you mobilize a whole other level and different kind of resources, including county governments, cities, and special interest groups, like the Indians and Friends of the Earth. If you do your planning on a participative basis, your planning leads to action." But, said Elmer, "if you do your planning on the outside with the Beltway Bandits, your planning leads to inaction. Gouraud's style is centralized planning with the help of the Beltway Bandits, and he's going to ask the states to endorse what he's done, and I predict disaster."

It will of course be a profitable disaster for the Beltway Bandits themselves, whose continued good graces with the federal solar energy program—at the usual expense of smaller concerns—were clearly reflected in contracts awarded by DOE in fiscal 1978. Among such contracts was a package of five dispensed in February to "stimulate production" of photovoltaic cells: ARCO Solar got $322,000; Sensor Technology (an ARCO subsidiary) $644,000; Solar Power Corporation (Exxon) $758,000; Motorola $677,000; and Solarex $559,000. Of those five, only the Rockville, Maryland firm of Solarex—though not a "small business" in the populist sense—is unconnected to the *Fortune 500* family of megacorporations. This pattern was repeated with the announcements in May and September of HUD Cycles 4 and 4-A in its residential solar "demonstration" program: out of $8 million awarded to 41 solar manufacturers, three national firms—Daystar (Exxon), Sunworks (ASARCO), and Revere (ASARCO)—alone received $4.3 million, or 54 percent of the total, and the top eight grantees, including Grumman, Solargenics (ARCO), and Solaron, received $5.3 million or 66 percent. (When HUD awards its Cycle 5 contracts in 1979, incidentally, the residential solar "demonstration" program will have run its mandated five-year course. Thereupon, with the *Fortune 500* in firm command of the national solar market, the program is scheduled to be terminated.)

Similar figures and percentages emerged from DOE's announcement in July of its own Cycle 3 awards for solar "demonstrations" in commercial facilities: Lennox/Honeywell, for example, received more than $1 million or roughly 10 percent of the $10.7 million consigned to 42 manufacturers. Other big winners in Cycle 3 were Northrup (ARCO), Sunworks (ASARCO), Solaron, Ying, PPG, Revere (ASARCO), and Daystar (Exxon). Meanwhile, according to *Solar Engineering* (December, 1978), the "giant Sears, Roebuck, and Company is now test marketing a domestic solar hot water system in six geographically diverse areas." Because of the company's "well known policy of a money-back guarantee," says the magazine, "the system is sold only on an installed basis with Sears technicians doing all of the installation work." At a cost of $1800 to $2000 per unit, the Sears solar system "is being marketed primarily through display ads in newspapers." The magazine notes that Montgomery Ward, according to a company spokesman, "doesn't feel 'the [solar] market is quite ready yet,' but he added that when the market is ready, we will be ready."

Despite presidential rhetoric, then, despite the enthusiasm of millions of Americans who celebrated Sun Day in May of 1978, despite the spewing of poisonous radiation from the crippled Harrisburg nuclear power plant in the spring of 1979—the corporate elite and its government proxies advance steadfastly down the hard energy path which they have affirmed again and again in strategies, speeches, and policy actions since 1971. This being the case, where is one to look for a glint of hope that the nation isn't doomed to suffocate, poison, or blow itself to pieces in the next 20 years? Is there anyone, anywhere in the country with the power, the courage, and the foresight to commence a serious transition toward the small-scale renewable technologies and decentralized energy grid that will be required for the nation's near-future well-being, indeed its survival? Is there a single federal official or institution working earnestly in that direction? Anyone in Congress or in top state government positions?

In Congress, as a matter of fact, particularly in the House of Representatives, there has evolved a coterie of elected officials and aides who have come to be known as the "Solar Coalition." Led by such figures as Representatives Richard Ottinger (New York), Elvin Baldus (Wisconsin), Neal Smith (Iowa), and Senators Gary Hart (Colorado) and Charles Percy

(Illinois), the "Solar Coalition" has been largely responsible for the basketful of well-intentioned but nonetheless inadequate solar legislation produced by Congress since 1976, including the Small (Solar) Business Assistance Act of 1978. There is, moreover, a sprinkling of bureaucrats in middle-echelon federal agencies who have braved the prevailing corporate tide to support the efforts of small-scale solar technologies to the extent permitted by bush-league budgets and political confinements. The most important of those agencies—the Community Services Administration—will be mentioned from time to time in the remainder of this chapter.

Among the virtues of the CSA and the "Solar Coalition" is their willingness to listen to the counsel of persons and private institutions who offer a knowledge and a sense of urgency regarding solar resource development badly needed at the federal level. One such institution is the Washington-based "Solar Lobby," formed in 1978 at about the time of the first annual National Solar Congress, which brought to Washington in August of that year more that 150 "solar activists" from around the nation to discuss priorities and tactics for hastening the arrival of a "Solar Age" in the United States. (Ironically, the National Solar Congress was funded by a grant from DOE, whose objective thereby, in addition to muting critics of its solar program, was to obtain information pertaining to "barriers and incentives" to solar development as viewed from state and local levels.)

The principal founder of the "Solar Lobby" and its companion organization, the Center for Renewable Resources, is Denis Hayes, author of *Rays of Hope* and originator both of national "Sun Day" in May, 1978 and of "Earth Day" in 1971. Six months after the National Solar Congress, Hayes and nine co-authors from the Solar Lobby and kindred organizations, including Common Cause and the Sierra Club, published a 40-page document entitled *Blueprint for a Solar America*. The stated intent of this treatise was to provide the federal government, or at least those government decision-makers who might be expected to listen, with a comprehensive body of rigorous policy alternatives to the more hamstrung conclusions and recommendations of the White House Domestic Policy Review, discussed above. The *Blueprint* is unequivocal in its demand that renewable energy resources be developed rapidly enough to supply a *minimum* of 25 percent of the nation's total energy needs by the year 2000.

This demand is elaborated by a raft of specific policy suggestions in categories ranging from "Consumer Protection" and "Financing" to "Competition," "Job Training," and "International Programs"—i.e., most of the categories addressed by the DPR. With regard to "Financing," for example, the *Blueprint* calls for additional federal subsidies aimed at putting solar "on a fair footing with the development of other energy forms" (an admittedly tall order, since those "other energy forms," according to the *Blueprint*, have received over $200 billion in federal subsidies, versus about $1 billion for solar). Allied proposals for capitalizing solar development include creation of a federal Solar Energy Development Bank, grants-in-aid to low-income purchasers of solar equipment, a $300-million program to solar retrofit public housing, and a 40-percent tax credit to businesses converting from fossil-fueled operations to "fuels from biomass."

Some of the *Blueprint's* most trenchant suggestions are made in relation to "Increasing Employment in Solar Development." Here, say the authors, the "federal government should adopt a policy that no future energy development program or major energy legislation may be approved unless a labor impact statement is first filed and considered [by Congress]." In addition, the federal government should jointly undertake with state and local governments a program to create community energy leagues that would generate jobs by hiring and training the unemployed to install solar devices, retrofit existing structures, conduct energy audits, and implement weatherization and conservation measures." A parallel suggestion calls upon DOE to "give states and localities financial and technical assistance to utilize alternative energy technologies in public facilities and to promote private use." The *Blueprint* specifies a budget of $25 million in 1980 "for programs to revise state and local codes" in favor of solar energy development.

At quite the other end of the administrative spectrum, the *Blueprint* endorses the establishment of a national "Solar Policy Council responsible for interagency coordination [between federal bureaus]," chaired by the Vice President [with] a small interagency staff." This potentially abusive locus of power would be monitored by a "Citizen's Solar Commission... charged with ensuring that all sectors of the public participate actively in solar energy policymaking." Finally, with respect to the "International Implications" of U.S. solar energy policy, the *Blueprint* declares that an early priority "is the reduction of the domestic cost of solar hardware as foreign markets expand. Economies of mass production in the manufacture, installation, and maintenance of solar equipment can be confidently predicted as cottage industries grow into significant economic entities." The better to service these "entities," the *Blueprint* suggests that the U.S. Export-Import Bank play a radical new role in "facilitating the global transition to renewable energy resources," including assistance to smaller American companies wishing to develop an export market for their solar goods and services.

These excerpts afford but a peek at the broader range of ideas and insights constituting the Solar Lobby's *Blueprint*, but they are sufficient to reveal both its strengths and its vulnerabilities. Chief among the former is the relative boldness of such demands as radical increases in federal solar spending ($50 billion by 2000), a law requiring "labor-impact studies" of energy legislation, and an activist role for the Ex-Im Bank in support of "small business." It is also a strength of the *Blueprint*—viewed pragmatically—that despite the novel character of some of its proposals, it is careful to frame them in a manner consistent with existing government policy categories, administrative procedures and structures, even language. Indeed, it is often difficult to distinguish the tone and terminology of the *Blueprint* from that of publications issued by DOE itself. This will assure it the widest possible audience among those officials and members of Congress for whom it is apparently intended.

On the other hand, if there is a serious conceptual weakness in the *Blueprint*, it is precisely its emphasis on actions to be taken by a centralized federal energy administration. Whether deliberately or not, the document leaves its reader with the impression that if solar energy isn't developed through sweeping new policies and programs at the federal level, then it won't be developed at all. This is depressing for two reasons: (1) such complete reliance on a centralized national government apparatus, even granting it the best of intentions, will necessarily result in programs and policies tending to favor larger institutions over smaller ones (this tendency is betrayed in the *Blueprint* itself through its reference to "cottage industries" growing into "significant economic entitites" which may then be assisted by the Ex-Im Bank in penetrating global markets); (2) it is difficult to believe that the innovative solar development policies recommended in the *Blueprint* will now be adopted by the very federal departments and officials who have striven since 1971 to avoid and even to subvert such policies on behalf of their corporate overseers. In short, the Solar Lobby's alternative to the White House "Domestic Policy Review" appears to suffer from many of the same inherent contradictions between noble policy objectives and ignoble political realities as the DPR itself.

These observations are by no means intended to detract unduly from the Solar Lobby's *Blueprint for a Solar America:* given the constraints within which the authors evidently felt they had to work, it is a valuable and stimulating document, reflecting enthusiasm, intelligence, and faith that the government eventually can be brought to do the right thing by its people. For many who are active in the small-scale solar community, however, the most appealing of the *Blueprint*'s suggestions are those which combine generous provisions for federal "seed-funding" and technical assistance with maximum local and regional autonomy regarding the character, timing, and management of the solar energy development programs to be pursued in each locale.

In fact, to return to the question asked earlier in this chapter, it is *only* in conjunction with local and regional solar development initiatives—after almost a decade of federal involvement in solar R&D—that anything close to the potential of small-scale applications developed by small producers for the commonweal of the community has been realized. Amory Lovins refers to this anomalous situation in his critique of the DPR in September, 1978: the report, he says, "implies that local and state action is following behind federal action. This is exactly backwards in most cases and gives a seriously misleading impression of where the action is." The government needs to learn, says Lovins, "that in a country this big and diverse, central management is much more part of the problem than part of the solution. The role of the federal government should be to facilitate where possible, to do the few things only that it can do that have to be done, and otherwise get out of the way."

Lovins' remarks have been documented by Ken Bossong, director of the Citizens' Energy Project quoted earlier, who in 1977 compiled a voluminous catalog of "Local/State Energy Projects" then underway in the U.S. It would be impossible even to summarize the scores of projects cited in the document, most of which relate to energy conservation and small-scale renewable resource applications, but here are a few examples: (1) a wood-energy conversion program in Vermont which has stimulated a six-fold increase in the use of wood for home-heating since 1970; (2) serious R&D programs by the governments of Iowa, Minnesota, California, and Montana designed to encourage small-scale renewable energy technologies specifically tailored to each respective regional climate (Montana's program is financed by a 2.5 percent levy on the state's coal-severance tax); (3) a "new town" project in Cerro Cordo, Oregon which hopes to achieve near-total energy self-sufficiency for its 2500 residents through the deployment of renewable technologies ranging from solar heating and methane gas production to horse, trolley, and bicycle transit systems; (4) a boldly innovative fuel conservation and solar development program in Davis, California, whose major components include a new building code strongly favoring solar applications, city-funded construction of prototype homes and apartments that are 90-percent solar powered, citywide gardening and waste-recycling systems, and heavy emphasis on bus and bicycle transportation— all aimed at reducing the city's annual fossil fuel consumption by 40 percent in 1990 (it has dropped five percent since 1975, while energy consumption in neighboring cities has risen six or seven percent).

"Ultimately," writes Bossong, "the success of national efforts to pull the U.S. out of its energy crisis will be determined within the communities across the country and not in Washington, D.C." He adds that no "federal energy plan could ever be comprehensive enough to allow for the differing energy needs of each region of the country." Bossong is alluding here to "the logic of the ecosystem" discussed at the opening of this chapter, and indeed, not only that principle but most of the others associated with community-based appropriate technologies are visible in the projects which he has compiled. In order to look closer at the operation of those principles in the light of actual experience, the remainder of this book is devoted to thumbnail case-studies of two very different but equally instructive community-based energy programs: (1) a solar greenhouse "outreach project" developed in New Mexico for what has proven to be nationwide application; (2) a courageous and fruitful attempt to penetrate the ghettos of New York City with the promise of appropriate technology.

The reader will observe, incidentally, that each of the community energy programs now to be examined manifests a body of specific characteristics which set it apart distinctly from the corporate-government solar operations discussed in previous chapters. One: while both alternative energy programs have been supported by federal tax-dollars, the bulk of that support has decidedly not come from the Department of Energy. Two: the federal government's "return on its investment" has been maximized to a rare extent in each of these programs, due chiefly to the emphasis on low-cost technologies and materials, systematic utilization of local resources, and the training of local people to continue the projects independently. Three: the "return on investment" has been further compounded by a "multiplier effect," where the energy models and examples established in the local programs have inspired duplication by other people and groups in the community. Four: each of the local energy programs has led directly to new employment and entrepreneurial opportunities for previously unemployed or under-employed community residents. Five: each of the programs has tended to unite an otherwise heterogeneous array of people and groups around technologies which are simple enough for everyone to get in on the act—including women, ethnic minorities, and elderly people who are conspicuously absent from the typical DOE project (most of which are strictly the preserve of white professional males).

The compound effect of these characteristics has been a richness of karma and generosity that seem to exude uniformly from each of the two community programs. This good karma is enhanced by the fact that most of the principal figures involved are motivated less by narrow financial and professional gain than by genuine love and concern for all human beings, their communities, and their fragile endangered biosphere. Too, the project principals seem to perceive a value and a dignity in work itself, almost for its own sake, or at least for its tendency to bring people together in the joy of mutual productive enterprise, which enables them to treat human labor not as a plague to be escaped through machines or hirelings but as still another "renewable energy resource" to be developed in harmony with the physical environment. It is in this sense, perhaps above all, that these hard-working appropriate technologists stand in contrast not only to the nation's corporate oligarchs but to many of those small solar energy "entrepreneurs" whose difficulties are profiled in earlier chapters.

With regard now to greenhouses: it is much to the discredit of the federal solar energy program—which has stuffed nearly $2 billion into the money-choked coffers of America's largest corporations since 1971—that most Americans in 1979, when asked if they would care to own a "solar greenhouse," would either stare blankly and scratch their heads, yawn with ennui, or chuckle at the question on their way to the supermarket for a 50-cent tomato. It is conversely much to the credit of people like Bill and Susan Yanda, solar pioneers from Santa Fe, that in just five years of disciplined, independent research and demonstration, the food-producing, heat-producing "attached solar greenhouse" has been established as one of the most efficient and in some ways most exciting alternative fuel technologies in the nation. Little thanks to DOE, moreover, the Yandas have been able to expose this marvelous device to as many as 100,000 people in New Mexico, Minnesota, Massachusetts, and a dozen or so other states since 1975.

An "attached solar greenhouse" is very much what its name implies—a basic greenhouse, generally built to retain a higher percentage of solar radiation than conventional greenhouses, attached to the south wall of a structure in such a fashion as to share its stored heat therewith. The variations on this theme are endless, from costly double-glazing to cheap polyurethane, from crystal garden palaces to humble truck farms, but the function is the same: to provide heat and moisture (if desired) to the "mother building," often assisted by a small electric blower, while also providing year-round vegetables in quantities limited chiefly by the owner's time and inclination. The Yandas specialize almost exclusively in the low-cost greenhouse, because the principal users of their technology—many of whom are elderly Mexican Indians on fixed incomes—happen to live in the poorer districts of the harsh New Mexico mountains around Santa Fe. In fact, "the logic of the ecosystem" which demanded thick adobe walls in this area 800 years ago is now a major factor in the development of modern passive energy concepts, including the solar greenhouse, for which the New Mexico solar energy community has grown rather famous. (Steve Baer and Steve Kenin, for example, both mentioned in Chapter V, are among the important passive solar technologists based in New Mexico.)

Having essentially perfected their greenhouse design by 1975, the Yandas first went noticeably public in 1976 through an "outreach program" called the Solar Sustenance Project, funded by the New Mexico State Energy Resources Board. In a published report on the project, Bill Yanda called it "the first state-supported program in the United States to take a partici-patory demonstration of solar energy directly into local com-munities." The object of the program was to build a series of 12 attached greenhouses on low-income dwellings in "widely diverse geographical and climatological areas" of the state. Each of the greenhouses, to be constructed by local participants in a two-day workshop, was intended to stress not only thermal aspects but also "the food and income capabilities" of the greenhouse concept—which featured such items as urethane "glazing" on a wooden frame with storage of heat in masonry foundations and water-filled 55-gallon drums.

"A great deal of the responsibility for the success of the workshops," writes Yanda, "was put into local hands." These "local hands," drawn from solar energy societies, Community Action Programs, and service organizations, were charged with locating a site for the greenhouse, ordering materials, planning a public meeting the night before the workshop, and publicizing the whole affair. They did well. Each public meeting—designed to "inspire the audience to come out and build the greenhouse the following two days"—was attended by an average of 75 persons from varying socioeconomic sectors (leading indirectly to the formation of three new solar energy societies), and the workshops themselves had an average participation rate of 90 persons, some who worked and some who watched. These community blitzkriegs left most of the greenhouses "75 to 98 percent complete" at the end of each workshop, with the final touches, including the planting of gardens, added later by the "owners of the greenhouses and interested friends."

It was then these "interested friends" and other parties who in the following year, according to a survey conducted by Yanda, raised no fewer than 300 additional greenhouses throughout the state—quite a handsome "multiplier effect." Doubtless one of the reasons for it was the unambiguous performance of the workshop greenhouses, which were monitored as part of the program. "I thought corn this sweet only grew in Iowa," said Rebecca Chavez to Yanda in reference to the late-April garden in her greenhouse. "On the average," writes Yanda, "though one person said her greenhouse (10 feet by 20 feet) produced all the (winter) vegetables a family of four plus guests could put away, each project greenhouse produced 40 to 60 percent of the family's vegetables throughout the winter. All participants have commented on the fantastic taste of a vegetable picked fresh in

January." The average total value of each family's greenhouse produce was $252.84 (1976 dollars), which was equal to 11 percent of the average family food bill that year.

In addition, writes Yanda, an "important aspect of greenhouse growing, which cannot be overstated, is using the greenhouse during the summer period." Summer night-time temperatures in the mountains of New Mexico, "often dropping into the low 50's, are hardly conducive to high production from fruiting vegetables. The well-run summer greenhouse not only avoids the ravages of hail and drought, but puts fruiting vegetables into an environment which can keep an optimum mid-60's low temperature range." Furthermore, says Yanda, the solar greenhouse "must be looked at in the light of its water conservation over field crop conditions. Authorities report water usage for greenhouse crops to be one-tenth to one-thirtieth of the field crop." Yanda adds that two of the project greenhouse owners were able to water their gardens year-round from "simple systems for trapping rain and melted snow from the roofs of their homes." He points out finally that because three-fourths of the cost of supermarket vegetables represents petroleum "used to transport, process, preserve and distribute them...a dollar's worth of December tomatoes or lettuce produced in the home greenhouse garden conserves about 75 cents' worth of petrochemical energy."

These crop returns from the project greenhouses were dramatically complemented by the heat they produced that winter of 1976. The "Griffin home" solar greenhouse, for example, though imperfectly oriented and partially shaded by trees, reduced the family's heating fuel costs as much as 37 percent during the winter months, with a total reduction of 24 percent for the year. When a Franklin wood-burning stove was added for back-up heat the following winter, the Griffins enjoyed monthly cost reductions as high as 52 percent and a yearly reduction of 40 percent. Parallel benefits of the greenhouse included its function as an "airlock" for the southward door and steel casement windows of the home, thus reducing winter-time heat losses due to convection and the opening and closing of doors. Moreover, writes Yanda, since "the greenhouse is used as a garden, it is providing the home with high-humidity air. The regular home furnaces and the wood stove are dry, forced air systems. The greenhouse is, in effect, providing the other heaters with air which has a greater capacity to store heat." (It cost the family a minuscule $127.55 to build this 120-square-foot solar greenhouse, because, says Yanda, "Griffin is a scrounger par excellence.")

Most of the other project greenhouses, except those without thermal storage, performed on a par with the Griffins', yielding an average heating-bill reduction of $94 per year, or about 30 percent of the owners' annual heating costs. This amount added to the savings in food expenses—$252.84—comes to a total of $350.84 in savings per family directly attributable to the project greenhouses. The average cost to build each greenhouse was $3.14 per square foot, or $512.40 for the "owner-built" greenhouse of 160 square feet. Simple arithmetic says that the "payback period" for these greenhouses—after which the owners will *net* a yearly cash-value return of $350.84 (1976 dollars)—is 1.4 years. Yanda observes that even the "contractor-built" solar greenhouses stimulated by the workshops, which cost an average of $1040 each, will be "paid back" in only 2.9 years. (The development of local "markets" for greenhouses built by contractors on the homes of persons too old or otherwise unable to build their own was a desirable and intended by-product of the Solar Sustenance Project. Another was the formation of local purchasing cooperatives, which lowered construction costs through bulk acquisition of materials.)

Not even the federal government could ignore the success and larger potential of the Yandas' greenhouse workshop program in New Mexico. To test its effectiveness in other U.S. communities, the Rural Development Service of the Department of Agriculture awarded the Yandas a contract in 1977 for a "Solar Greenhouse Outreach Program" to be conducted in five very different regional climates in Massachusetts, Virginia, Kentucky, North Carolina, and Arkansas. The format of the outreach program was modeled closely on that of the New Mexico workshops—an evening public lecture followed by a two-day "participatory" workshop comprising the design and construction of a low-cost solar greenhouse—with heavy involvement by local coordinators at each of the targeted sites. Again drawn chiefly from community action agencies, these local coordinators were expected not only to organize and publicize the workshops but also to raise the money for greenhouse construction materials.

In their final report on the five-state program, the Yandas emphasize the "process" and "problem-solving" aspects of the workshops, as well as the importance of community participation and leadership. "If the structure of the workshop is not closely defined or limited," write the Yandas, "exciting possibilities for teaching and learning evolve." While it would be simpler to construct each greenhouse from "a professional quality blueprint," that would be "counter-productive to the overall goals of the workshop. Using the workshop as a *process* demands a different viewpoint and managing role by the project leaders." The Yandas contend that each "major building decision must be open to explanation and group criticism. This means the training team should be prepared to explain and/or defend a particular construction technique or design against opposing views and, if necessary, change or compromise on the final product."

Indeed, at the workshop in Charlottesville, Virginia, the training team was forced by local geological conditions to improvise a new technique for building the greenhouse foundation— "below-grade pressure-treated wood panels," instead of the usual concrete block or slab— a collective solution by the greenhouse workshop which likely would not have emerged except through the input of local participants. "An ability by the workshop leaders to read local sensitivities," write the Yandas, "can often determine the success or failure of the project." The Yandas also stress, both in their report and in their workshop programs, that "the final cost of a greenhouse is mainly a function of the prospective owner's aesthetics and not necessarily an accurate indicator of performance. An owner with the time commitment to spend in gathering recyclable materials, building the greenhouse, manually moving insulating panels, and carefully tending the plants can have as successful and productive a greenhouse as an expensive automated unit. A revolutionary aspect of solar greenhouse implementation," add the Yandas, "is that high performance can be attained through human involvement." This fact among others makes the solar greenhouse "available to all levels on the economic scale."

Problems and occasional crises notwithstanding, the Yandas' five-state "Outreach Program" repeated the success of their efforts in New Mexico. The workshop in Hyannis, Massachusetts, for example, led to a contract from the state Department of Community Affairs to fund the construction of 10 additional solar greenhouses by the "New Resource Group, Inc., a local private firm whose personnel were trained at the (Yandas') Hyannis Cape Workshop." In Charlottesville, an elderly couple named Sprouse are now deriving up to 40 percent of their winter heat from the solar greenhouse attached to their isolated mountain cabin—a special boon in their case, since Mr. Sprouse, who used to chop wood for the couple's heat, was partially disabled by a stroke not long after his greenhouse was completed. The project greenhouse in Murray, Kentucky generates mid-winter daytime temperatures of 80 to 90 degrees Fahrenheit, despite the initial skepticism of local area college professors who observed the design and construction process. Finally, at the workshop in Durham, North Carolina, a little army of 66 persons—young and old, black and white, male and female—signed up to build the greenhouse, on completion of which the training team was besieged with requests for greenhouse workshops by other Durham community service organizations.

Success begets success: the Yandas accepted a contract in 1978 from the New Mexico Energy Extension Service (funded by DOE) to train a series of three-person teams from 20 different states in the skills required "to organize and manage solar greenhouse workshops." Each team, write the Yandas, "will contain one person with building/design experience, one with horticultural expertise and one with community mobilization skills. Hence, the three most important elements of the successful workshop are embodied in one team," whose training will focus the team's "expertise where it belongs, at the local level." The Yandas hope that an eventual proliferation of such teams in local communities across the United States will help correct the mistaken public view "that solar energy is fine for the south or southwest but cannot pay back in areas of the country with less sunshine or colder weather. This view," say the Yandas, "still held by a majority of the citizenry, is possibly a result of inaccurate national media presentation which has given emphasis to very expensive solar systems."

Most of the nation's small-scale solar development efforts, including Bill and Susan Yanda's, can roughly be grouped into three separate categories defined in terms of the "users" of a given solar system or technology: 1) middle class urban and suburban homeowners in large to medium-size American cities (e.g., Davis, California); 2) moderate to low-income "marginal" families, comprising both renters and homeowners, in smaller towns and quasi-rural areas (e.g., Crystal City, Texas, and Portales, New Mexico); 3) working farm families at a variety of income and "lifestyle" levels in clearly rural communities. These three "user" categories have received the bulk of on-site energy development attention for the simple reason that most existing solar, wind, and biomass technologies are more readily deployed in such contexts. The factors operating in favor of these contexts range from private home ownership and relative purchasing power to open space, accessibility of construction materials and natural resources, community cohesiveness, and a psychological willingness to experiment with new ideas.

Even under the best of circumstances, where, say, *all* of the above characteristics obtain, the implementation of on-site renewable energy technologies has been a spotty and tedious enterprise, distinguished more by its heartbreaks than its breakthroughs. Where, on the other hand, virtually *none* of the above characteristics obtain, as in the ghettos of large northeastern American cities, any group attempting to introduce small-scale solar energy ideas will be a minority within the minority of small-scale solar proponents in general.

It is just such a group which has thrown down the gauntlet for appropriate technology in the least hospitable arena for such a struggle in the United States: the tenement districts of New York City, from the rubble-strewn highlands of the South Bronx to the teeming narrow streets and flame-charred walkups of the Lower East Side. This group—the Manhattan-based Energy Task Force—was organized in 1975 by a handful of architects, engineers, and educator-activists who believe that if poor people are to survive in an urban context they will have to impose community control over elemental life-support systems, including housing, energy supply, and food production. Accordingly, and quite unmindful of "profit," ETF has joined forces with a *potpourri* of neighborhood cooperatives which themselves are part of a larger urban movement toward "squatter" occupation and "sweat-equity" salvage of inner-city ghetto buildings abandoned and often torched—for insurance purposes—by their absentee owners.

One such "squatter" cooperative is a group called "Charas," composed primarily of black and Puerto Rican families who focused their initial survival efforts—they have since expanded to five buildings—on a five-floor walkup at 519 East 11th Street on the Lower East Side (*Loisaida*) of Manhattan. In 1973, during a rent strike provoked by the landlord's refusal to extend basic maintenance services, 13 mysterious fires were ignited in the building—despite surveillance by tenant patrols—leaving it finally a gutted shell. Yet the tenants refused to move. Instead, when the city seized the building with demolition in mind, "Charas" was born and purchased the building for a token fee of $2000. This was followed by a "sweat-equity" loan which enabled the tenants to renovate the building themselves, reoccupying it in March, 1976. The result of these arduous maneuvers was a tenant-owned, structurally sound building of 11 apartments whose price-per-room had been reduced from $110 monthly under the original owner to $55 under cooperative management.

The inevitable stinger in an otherwise happy ending was New York City's brutal energy costs, particularly electrical energy at 10 cents per kilowatt hour, but also fuel oil, which has tripled in price since 1971. Further escalations might well have bankrupted "Charas" in two or three years, and it was here that the Energy Task Force entered the picture. ETF "was drawn to this devastated low-income neighborhood," says a published Task Force report, "because of the group's dedication to illustrating the crucial relationship between energy-efficient design and affordable housing." ETF saw a chance to "demonstrate to low-income people that they could help themselves with energy-saving and energy-producing efforts." (It is further noted in the Task Force report that while "nearly 30,000 housing units are abandoned every year in New York City...municipal and federal housing efforts had been failures on the Lower East Side until the self-help housing movement began to develop within the community.")

Having learned of "Charas" through a television talk-show featuring one of its members, ETF worked closely with the cooperative on a series of energy development projects in 1976 and 1977 which now stand as models of small-scale solar technology applied in an urban milieu. The guiding principle in each of these projects, as in the bulk of ETF's work in low-income communities, was maximum reliance on the skills, labor, and other resources of the community itself. This has meant, for the most part, that ETF professionals have functioned as designers, teachers, and consultants, while co-op members have functioned as apprentices, construction supervisors, and installers.

Phase one of the work at 519 East 11th Street was to weatherize the building, a task accomplished by ETF engineers and building tenants during the latter stages of renovation early in 1976. Phase two was the installation of a roof-top solar hot-water system engineered by ETF and "Charas" with solar panels produced by a small minority manufacturer in Brooklyn. Phase three involved the design, assembly, and erection of a two-kilowatt wind-electric machine atop a 37-foot tower on the roof of the building—not three ghetto blocks from the smoke-belching stacks of a giant Con Edison electric plant near the putrid shores of a fishless East River.

The decision by "Charas" to put up a windmill was rather an afterthought, spurred by the fact that Con Ed had discontinued electric service to the co-op because of a billing dispute. On that very day, as "Charas" members and ETF staffers con-

ferred in the building on East 11th Street, they received a chance
visitor in the person of Ted Finch, a recent master's graduate in
wind-electric technology from MIT who had come to Manhattan
to deliver a series of lectures on the potential of New York's
harbor winds for electric power generation. Finch was instantly
recruited by ETF, whose staff until then had not included such a
specialist, and planning commenced on the spot for the "Charas"
wind-electric machine. When completed in November, 1976—
again through the efforts and accumulating skills of "Charas"
members themselves—its principal component was a rebuilt
Jacobs Wind Charger (circa 1925) which produces electric power
for the pumps and motors in the co-op's solar system, as well as
for common lighting in corridors and stairwells.

Con Ed, meanwhile, didn't take kindly to the sudden
appearance of the wind machine, and less kindly still to a
proposal by "Charas" and ETF that the cooperative feed any
excess power back into utility lines, thereby earning credit
against the company's monthly charges to the co-op itself. Such
an arrangement, said Con Ed, according to *The New York Times*
(May 6, 1977), might cause "power surges" that would damage
the monopoly's 10-million-kilowatt transmission system. It took
a special ruling by the New York State Public Service Commission to change the utility's mind, and even then Con Ed was
allowed to invoke a monthly "surcharge" against the cooperative
that demolished what little credit might otherwise have accrued.
Still, the decision by the P.S.C. "in favor" of "Charas" was
celebrated by a party at the base of the windmill which attracted
people from all over the neighborhood. Some of them, for the
benefit of a *Times* photographer, climbed the wind-tower and
clenched their fists in the traditional symbol of revolutionary
struggle.

That, of course, is the point. Con Ed must know that its "surcharge" to the tenants at 519 East 11th Street is a trifling matter in
relation to the value of that little windmill gyrating day and
night in naked view of thousands of indigent victims of corporate America. The value of the wind-machine, like the value
of the "Charas" solar collectors tilted toward the sky, is
symbolic and inspirational. It says to the people in the streets

below that a different life is available to them—a life of dignity and independence, of self-reliance and reward for honest, self-managed work. "The crippling fear of bureaucracy which so often frustrates the poor is absent on East 11th Street," says the ETF report on the project: "People there have moved from squatters' instincts to organized, long-range planning for economic growth through sweat-equity construction and energy conservation programs. The windmill at 519 represents just a small part of the success they've made at self-help."

While there is no way to attach a dollar-figure to such results, it is worth noting that even in dollar-terms the "Charas"/ETF energy projects—funded by the Community Services Administration—represent a wholly more prudent use of the taxpayers' money than DOE's power-towers and thrice-over contracts for studies of studies of OTEC plants. The $60,000 invested by "Charas"/ETF to weatherize and solar-equip the building on 11th Street, for example, will be "paid back" through reduced utility bills in less than seven years. That is also the estimated "payback period" for the $4500 Jacobs wind-machine, despite a daily kilowatt yield that is lower than expected due to lower average wind velocities. It would seem, as an ETF engineer has said, that "coughing up a grant for solar collectors and windmills is a much more efficient way to supplement low incomes than welfare checks."

The abovementioned correlation between energy self-sufficiency and a new sense of strength and dignity for the poor has likewise informed the numerous other "self-help" enterprises which ETF has initiated or supported in the barrios of New York. Among the few which can be summarized here is the famous "CUANDO Solar Wall," constructed in the summer of 1978 by ETF activists and members of a controversial Lower East Side youth organization called *Culturas Unidas Aspiran Nuestro Destino Original*. (This loses a little in its English version: "Cultural Understanding and Neighborhood Development Organization," but it still reduces to CUANDO, meaning "when.") Part of the significance of the CUANDO Solar Wall, which stretches across the entire south face of CUANDO's headquarters between its second and third floors, is the fact that it is a *passive* solar device requiring no moving parts, no electrical power, and virtually no maintenance. ETF's Ted Finch, who helped design and construct the wall, says that he has come to feel "really troubled by the level of sophistication necessary to build and maintain active solar systems" (of the sort generally specified, for example, in HUD/DOE "demonstration" contracts). "If low-income people with no technical skills are to get a jump on the cold," says Finch, "solar systems have to be kept simpler and maintenance kept to a minimum."

This describes the CUANDO Solar Wall, whose 500 square feet of fiberglass collector, mounted flush on the side of the building, now produce heat for a community gymnasium which before was so frigid in December and January that it could scarcely be used. (In summer the function of the wall is reversed to help cool the interior of the building.) "'The beauty of the passive system,'" says Finch, quoted by *In These Times* (Oct. 11-17, 1978), "'is that aside from opening a few ducts to regulate the flow of air heated in the solar wall, the whole thing runs itself.'" Because of the passive features of the wall, it was much easier to involve the young people of CUANDO in its actual design and construction. "'We didn't want anything that we couldn't handle ourselves,'" says Richard Cleghourne, program coordinator for CUANDO who was also quoted by *In These Times*: "We can't have something that would be too much trouble or too expensive." Indeed, the wall was co-designed by CUANDO's Fred Cabrera and constructed by *Loisaida* teenagers during a summer seminar on "energy alternatives" taught by Finch. "CUANDO is out to prove the point," says *In These Times*, "that solar energy is clearly a feasible alternative for poor people to the expensive and uncontrollable costs of commercial energy."

Here is a sampling of still other projects in which the small staff of the Energy Task Force—from five to eight at any given time—has been substantially involved: 1) planning, design, and user-assisted construction—with funding support from CSA—of a 40-kilowatt wind-turbine for a group called the Bronx Frontier which uses the wind-power to aerate compost for a very ambitious community gardening project in the urban desolation of Hunt's Point; 2) a solar hot-water system for a building owned by the South Bronx People's Development Corporation; 3) a 12-week "energy auditor" training program in Harlem designed to equip unemployed Harlemites with the skills required to analyze and improve the energy-efficiency of tenant-owned buildings in the community; 4) a one-year "Solar Utilization and Employment Development" program, jointly funded by the U.S. Department of Labor, CSA, and DOE, whose objective is to furnish on-the-job training in solar design and installation techniques for 15 underemployed inner-city workers, mainly black and Puerto Rican.

One would think, in view of its success rate and productivity, that a group like the Energy Task Force would be assured a long and well-endowed future by a grateful federal government. As of March, 1979, however, the survival of the group was being threatened by the same bureaucratic machinations that have crippled the rest of the U.S. solar energy program and periodically hampered the operations of ETF itself. This threat has emerged in the form of a tentative decision by the White House Office of Management and Budget, which funds and oversees the Community Services Administration, to reduce that agency's budget for solar programs from $65 million in FY 1979 to $10 million in FY1980. The ostensible reason for this decision, according to ETF's Chip Tabor, is twofold. "Number one," says Tabor, "this is Jimmy Carter's idea of fighting inflation. Number two, Schlesinger wants to concentrate all solar energy programs in the Department of Energy. But," Tabor hastens to add, "everybody knows that DOE is hardware-oriented. The only solar projects that are people-oriented have come out of CSA. This is a slap in the face to groups like us. It will seriously damage our work in the community."

Should ETF in fact be weakened or destroyed by this development, it would be difficult to escape the conclusion that, in the eyes of certain federal officials, the group has perhaps been *too* effective on behalf of its potentially militant constituents. "When people solve one problem, like energy, on the fringes of their lives," said ETF's Margaret Morgan in an interview

view with *Newsday* (Oct. 8, 1978), " they get more determined to solve other problems. It doesn't matter where you start, as long as the effort gives you more determination and more control. " This was echoed in the *Newsday* article by Larry Levan, 22, from Harlem, who was trained by ETF to conduct energy audits for The Renegades Housing Movement, Inc. " This is a start, " said Levan, " for communities to get back control over their own existence. "

It is easy to imagine the crystals of sweat popping out on the brow of a former Lockheed vice-president—or a former director of the CIA—who in his tenure at the Department of Energy is suddenly made aware that ghetto groups with names like "The Renegades" and "CUANDO" are cheerfully using federal solar energy funds to build a mass movement toward community independence and self-determination: wasn't that the purpose, just ten years ago, of the Black Panther Party and the Students for a Democratic Society? The tenant farmers of Mississippi and Alabama during the Depression? The democratic rebels of the IWW who seized Seattle in 1917? Where, indeed, might such a movement lead through the turbulence of the 1980's? Once "The Renegades" have secured even partial control over such of their staples as energy, housing, and food— once they have severed the tape and wire that have held them in thrall to Exxon, Con Ed, Safeway, and HEW—what is to keep them from reaching for more? Suppose they should come to presume, in concert with like-minded people in other communities across the land, that they are capable of running their own schools and transportation systems, their own factories, shipyards, and copper mines, their own TV stations, newspapers, even banks? What then? Could the citadels of government be far behind?

EPIGRAMS

1. Denis Hayes, *Rays of Hope* (New York: Norton, 1977).

2. Piet Bos, "Solar Realities," *EPRI Journal*, Electric Power Research Institute, 3412 Hillview Avenue, Palo Alto, California 94304 (February 1976).

3. *Solar Energy: Progress and Promise*, Council on Environmental Quality, Washington, D.C. (April 1978).

4. The McCormack quote is taken from a speech transcribed in *Energy Awareness: Symposium for Public Awareness on Energy* (held in February 1976 in Knoxville, Tennessee), published by the U.S. Energy Research and Development Administration, Washington, D.C. (1976); see Chapter IX for an elaboration of this speech and others at the symposium.

CHAPTER I

1. Material on the Fisks' experience in Crystal City compiled from interviews, private communications, and documents published by the Center for Maximum Potential Building Systems, 8604 FM 969, Austin, Texas 78724.

2. *The National Energy Plan*, Executive Office of the President, Energy Policy and Planning, White House, Washington, D.C. (April 1977).

3. Commoner's remarks are from a speech at the University of Texas, Austin (February 4, 1978).

4. The studies in 1974 by TRW, General Electric, and Westinghouse are available from the National Science Foundation, Washington, D.C.

5. Figures on solar contracts to small business in 1976 are from "Large Corporations May Soon Own the Sun," *People & Energy*, 1757 "S" St. NW, Washington, D.C. (May 1976).

6. "Energy Research and Development and Small Business," *Hearings of the Select Committee on Small Business*, U.S. Senate (May 13 and 14, October 8 and 22, November 18, 1975).

7. Amory B. Lovins, "Energy Strategy: The Road Not Taken?" *Foreign Affairs*, vol. 55, no. 1 (October 1976); see also Lovins, *Soft Energy Paths: Toward a Durable Peace* (Cambridge, Massachusetts: Ballinger/Friends of the Earth, 1977).

8. Robert H. Murray and Paul A. La Violette, *Assessing the Solar Transition*, International Center for Integrative Studies (U.S. office), 45 West 8th St., New York, New York 10011 (1977).

9. Robert Engler, *The Brotherhood of Oil* (Chicago: University of Chicago Press, 1977).

10. John E. Tilton, *U.S. Energy R&D Policy: The Role of Economics*, NSF and Resources for the Future, 1755 Massachusetts Ave., NW, Washington, D.C. (1974).

11. Schlesinger's speech reprinted in *Fortune* (February 1976).

12. "New Energy Department: Where It Goes From Here," *Nation's Business* (October 1977).

13. Schlesinger's remarks in Brussels reported by Associated Press in *American-Statesman*, Austin, Texas (October 9, 1977).

CHAPTER II

1. Richard Barnet and Ronald Müller, *Global Reach: The Power of the Multinational Corporations* (New York: Simon and Schuster, 1974).

2. Engler, *op. cit.*

3. *United States Energy Outlook*, National Petroleum Council, Committee on U.S. Energy Outlook, Washington, D.C. (1972).

4. *A Time To Choose*, Energy Project of the Ford Foundation (Cambridge: Ballinger, 1974).

5. *Nuclear Power, Issues and Choices*, summarized in *Energy Daily*, 300 National Press Building, Washington, D.C. 20045 (March 22, 1977).

6. *Solar Energy as a National Energy Resource*, National Science Foundation and National Aeronautics and Space Administration, Washington, D.C. (1972).

7. Sandra Oddo, "An Interview with Karl W. Boer," *Solar Age*, P.O. Box 4934, Manchester, New Hampshire 03108 (June 1978).

8. Information on ERDA solar contracts in 1976 and 1977 compiled from ERDA (Solar Division) "Program Summaries" and "Mission Analysis" documents for those years, as well as from an unpublished list of "Active Prime Contracts, Federal Solar Program" (July 1977).

9. Harry C. Boyte, "Roundtable Wields Immense Power Behind the Scene," *In These Times* (September 14-20, 1977).

10. Engler, *op. cit.*

11. Information on EPRI compiled from *EPRI Journal,* Electric Power Research Institute, 3412 Hillview Ave., Palo Alto, Ca. 94394.

12. *Annual Report,* Resources for the Future, *op. cit.* (1976).

13. *The Nation's Energy Future,* Atomic Energy Commission, Washington, D.C. (December 1, 1973).

14. Simon's remarks are from an interview in *Science* (April 19, 1974).

15. Tilton, *U.S. Energy Policy: The Role of Economics, op. cit.*

CHAPTER III

1. Engler, *op. cit.*

2. *Federal Energy Advisory Committees,* U.S. Senate, Washington, D.C. (1977).

3. "Energy Research Advisory Board," roster of members from U.S. Department of Energy, November 1977.

4. Information on ERDA General Advisory Committee and "Solar Working Group" compiled from resumes supplied by ERDA (July 26, 1977).

5. Recent information on Frank Zarb compiled from *Energy User News* (September 4, 1978), *Energy Daily* (June 28, 1978), *Time* (January 22, 1979).

6. Boyte, *op. cit.*

CHAPTER IV

1. *Public Utilities Fortnightly,* Public Utilities Reports, Suite 500, 1828 "L" St., Washington, D.C. 20036 (September 26, 1976).

2. Additonal information on Seamans from *Energy Daily* (May 1, 1979).

3. Common Cause, *Serving Two Masters,* a study of corporate-government interlocks, Washington, D.C. (1976); see also Angus McDonald, "Solar Energy in Washington," a research paper presented to the Midwest Electric Consumers Association (July 1976).

4. *Annual Report,* The MITRE Corporation, Box 208, Bedford, Massachusetts 01730 (1976).

5. Information on federal solar contracts to MITRE and other corporations drawn from ERDA/DOE "Program Summaries" and "Mission Analysis" documents (January 1975-March 1979).

6. Telephone interview with Don Elmer, Geothermal Division, U.S. Department of Energy November 1977.

7. *Science and Government Report*, 3736 Kanawha St., Washington, D.C. 20015 (April 1, 1976).

8. Information on Henry Marvin from *Solar Energy Washington Report* (August 11, 1975).

9. Howard Coleman quoted in *Environmental Action Bulletin*, Rodale Press, Inc., Emmaus, Pennsylvania 18049 (August 20, 1977).

10. Private communication from Robert J. King, Texas Energy Advisory Council, November 1977.

11. "Energy Conservation's Impact on R&D," an extensive report in *Business Week* (June 27, 1977).

12. Information on J. Hilbert Anderson compiled from telephone interviews (August 1977 and February 1979), ERDA/DOE "OTEC Program Summaries" (1976 and 1977), and an unpublished research paper by Ernest W. Harkins, Austin, Texas (November 1977).

13. OTEC contract awards compiled from ERDA/DOE "OTEC Program Summary" (1976 and 1978) and *Energy Daily* (May 2, 1978).

CHAPTER V

1. *Hearings of the Select Committee on Small Business, op. cit.*

2. Private communication from Steve Kenin, The Solar Room Company, Box 1377, Taos, New Mexico 87571.

3. HUD "Cycle 2" awards for residential solar demonstrations, U.S. Department of Housing and Urban Development, Washington, D.C. (June 1975).

4. Information on Primack solar homes from "The Growing Rush to Solar Energy," *U.S. News and World Report* (April 4, 1977).

CHAPTER VI

1. *The Future of Small Business in America*, Subcommittee on Antitrust, Consumers, and Employment, Committee on Small Business, U.S. House of Representatives, Washington, D.C. (November 1978).

2. *Solar Engineering*, 8435 N. Stemmons Freeway, Dallas, Texas 75247 (June 1978).

3. Information on $5-million appropriation for Small (Solar) Business Assistance Act of 1978 furnished in a private communication from Jim Piper, Piper Hydro, Inc., 2895 E. La Palma, Anaheim, California 92806 (February 1979).

4. Telephone interview with Jim Piper (September 1977).

5. HUD "Cycle 3" solar demonstration awards, U.S. Department of Housing and Urban Development, Washington, D.C. (June 1977).

6. "Second Round" of commercial solar demonstration contracts, ERDA (DOE), Washington, D.C. (March 1977).

7. Marvin quoted in *Energy Daily* (May 11, 1977).

8. Information on 1976 wind-electric contracts are from *Solar Engineering* (October 1976) and *The New York Times* (June 27, 1976).

9. Allen L. Hammond and William D. Metz, "Solar Energy Research: Making Solar After the Nuclear Model," *Science* (July 15, 1977).

10. "The Trouble at Stillwater: ERDA Rejects Small Wind Systems," *Windustries*, Box 126, Lawrence, Kansas 66044 (Winter 1976).

11. "Federal Wind Energy Program Summary," DOE (January 1978).

12. The five-year budget projections by the General Accounting Office are summarized in *Energy Daily* (February 23, 1978).

13. Lovins, *op. cit.*

14. William D. Metz, "Solar Thermal Electricity: Power Tower Dominates Research," *Science* (July 22, 1977).

15. Information on Solar Dynamics, Ltd. compiled from interviews with and documents furnished by Dave Marke, Chief of Research for Solar Dynamics, Ltd., 3904-C Warehouse Row, Austin, Texas 78704 (1977-1979).

16. Cynical remarks about ERDA from a telephone interview with Peter Hunt, aide to former U.S. Senator Howard Metzenbaum (November 1977).

CHAPTER VII

1. The letter to Kauffman from Hart and Ottinger (October 29, 1976) and Kauffman's response (November 9, 1976), with an example of the offending Exxon ad, furnished by the office of Senator Gary

Hart, U.S. Senate, Washington, D.C. 20510 (May 9, 1977); see also Edward Cowan, "Who Should Develop Solar Energy?" *The New Hill*, 1970).

2. "Solar Energy" in *Shell Reports*, Shell Oil Company, Public Affairs, Room 1541, P.O. Box 2463, Houston, Texas 77001 (January 1976).

3. Piper's remark on "payback" is from "Sun Wars," Stephen Singular in *New Times* (September 30, 1977).

4. "Energy Is Your Business," visual communications kit from U.S. Chamber of Commerce (Attn: Robert Moxley), 1615 "H" St. NW, Washington, D.C. 20062.

5. "Our Energy Problems and Solutions," Gulf Oil Corporation, Consumer Affairs Division, P.O. Box 1563, Houston, Texas 77001.

6. Letter to Rhett Turnipseed from Jon M. Veigel, California Energy Resources Conservation and Development Commission, 1111 Howe Ave., Sacramento, California 95825 (June 24, 1977); memo from Jeff Reiss and Gary Starr to Veigel (June 24, 1977).

7. Lee Johnson, "ERDA—Stonewalling Small Business and the Sun," *Rain*, Eugene, Oregon (October 1977).

8. Telephone interview with Jim Benson, now director of the Institute for Ecological Policies, 9208 Christopher St., Fairfax, Virginia 22031 (November 1977).

9. *Definition Report: National Solar Energy Research, Development, and Demonstration Program (ERDA-49)*, U.S. Government Printing Office, Washington, D.C. (1975).

10. *Solar Energy in America's Future*, prepared by Stanford Research Institute (Palo Alto, California) for ERDA (DOE), Washington, D.C. 20036 (March 1977).

11. "The International Energy Situation: Outlook to 1985," U.S. Central Intelligence Agency (April 1977).

12. Stansfield Turner's congressional testimony is summarized in *Environment* (June/July 1977).

13. "Activities of the Office of Energy Information and Analysis," congressional task force investigation reported in the *American-Statesman*, Austin, Texas (December 23, 1977).

CHAPTER VIII

1. Sheldon Novick, "The Electric Power Industry," *Environment* (November 1975).

2. Lovins, *op. cit.*

3. "Memorandum of Understanding Between Electric Power Research Institute, Inc. and Energy Research and Development Administration" (May 25, 1976) and "Definition and Explanation of MOU and IA" furnished by Bob Ritzman, ERDA (DOE), Washington, D.C. (August 1977).

4. Telephone interview concerning the MOU with Bob Ritzman, ERDA (August 22, 1977).

5. "EPRI and ERDA Sign R&D Agreement," *EPRI Journal* (July/August 1977).

6. "Joseph Fisher: An Energy Leader in Congress," *EPRI Journal* (April 1977).

7. *Nuclear Power, Issues and Choices, op. cit.*

8. "A Round of Response," *EPRI Journal* (January/February 1978).

9. Barnet and Müller, *op. cit.*

10. Zbigniew Brzezinski, *Between Two Ages* (New York: McGraw-Hill (1970).

11. Chauncey Starr, "A Strategy for Electric Power," *Energy Awareness: A Symposium for Public Awareness on Energy, op. cit.*

12. Postma's remarks are from *Energy Awareness, ibid.*

13. James S. Cannon and S.W. Herman, *Energy Futures: Industry and the New Technologies* (New York: INFORM, 1976).

14. Rowland Evans and Robert Novak, "Environmental Backfire," *American-Statesman, op. cit.* (January 19, 1978).

15. Commoner at the University of Texas, *op. cit.*

16. Lovins, *op. cit.*

17. "ERDA/AGA Coal Gasification Pilot Plant Program," a summary report by the American Gas Association, 1515 Wilson Blvd., Arlington, Virginia 22209 (1976).

18. Engler, *op. cit.*

19. *EPRI Journal* (February 1977).

20. *The Nation's Energy Future, op. cit.*

21. *An Analysis of Federal Incentives Used To Stimulate Energy Production*, Battelle Pacific Northwest Laboratory, Richland, Washington (November 1978).

22. Lovins, *op. cit.*

CHAPTER IX

1. Except where noted, all information for this chapter is taken from *Energy Awareness: A Symposium for Public Awareness on Energy, op. cit.*

2. "GAO Hits ERDA Role in California Nuclear Debate," *Science and Government Report* (October 15, 1976).

3. La Grone's promotion reported in *Energy Insider*, house publication of U.S. Department of Energy, Room 8E070, Washington, D.C. 20585 (August 21, 1978).

CHAPTER X

1. Richard Balzhiser, "Coal: The R&D Pivot for New Energy Fuels, Cycles, and Storage," *EPRI Journal* (January/February 1977).

2. *The Cost of Energy from Utility-Owned Solar-Electric Systems*, Jet Propulsion Laboratory, California Institute of Technology, Pasadena, California (June 1976).

3. Information on EPRI-funded solar projects from *EPRI Journal* (January/February 1977).

4. W. Donham Crawford, "The Electric Utility Industry Looks at the Carter Program," *Aware*, Community Performance Publications, Inc., 2038 Pennsylvania Ave., Suite 12, Madison, Wisconsin 53704 (May 1977).

5. Information on WEEI newscast conveyed in a private communication from Robert Charlton, Solectro Thermo, Inc., 1934 Lakeview Ave., Dracut, Massachusetts 01826 (March 1979).

6. "Questions and Answers About Nuclear Power," *The Rural Electric Missourian*, Jefferson City, Missouri (July 1977).

7. *Electric Utility Solar Energy Activities*, Electric Power Research Institute, Palo Alto, California (February 1978).

8. Telephone interview with Christine B. Sullivan, Secretary of Consumer Affairs, The Commonwealth of Massachusetts, One Ashburton Place, Boston, Massachusetts 02108 (September 1977).

9. Telephone interview with John F. Meeker, New England Electric System, 20 Turnpike Road, Westborough, Massachusetts 01581 (September 1977).

10. "Interim Report on the New England Electric Solar Water Heating Experiment," prepared by Arthur D. Little, Inc. (Boston) for New England Electric System (May 1977).

11. The ERDA report on utility/solar options is summarized in *Solar Outlook*, Observer Publishing Co., 1054 31st St. NW, Washington, D.C. 20007 (April 4, 1977).

12. "Project S.A.G.E." and "Operation Sunflower" are discussed in "Utilities Eclipse Sun" by Jim Rosapepe in *Politicks and Other Human Interests* (October 25, 1977); see also Fred Branfman, "California's Fight for the Sun," *The Nation* (June 18, 1977).

13. Peter Barnes, "Utility vs. Consumer Ownership of On-Site Solar Energy Systems," The Solar Center, 944 Market St., Suite 320, San Francisco, California 94102 (April 14, 1977).

14. "The Solar Energy/Utility Interface: Report on Workshops and Conference," The Office of Energy Programs, School of Engineering and Applied Science, George Washington University, Washington, D.C. 20052, prepared for the U.S. Department of Energy (October 1, 1978).

CHAPTER XI

1. Barnet and Müller, *op. cit.*

2. "U.S. Giants Stepping into Solar Industry," *Solar Engineering* (December 1976).

3. "Will Factory-Built Solar Homes Corner the Market?" in *Solar Energy Digest*, P.O. Box 1776, San Diego, California 92117 (August 1977).

4. "Energy Conservation's Impact on R&D," *Business Week, op. cit.*

5. Information on Steve Kenin from a private communication, *op. cit.*

6. Information on Sunworks/ASARCO from a variety of sources, including periodic "Program Summaries" and contract announcements by HUD and DOE; see also "Business Beat" in *Solar Age* (November 1976).

7. *The Fight for the Sun*, unpublished manuscript by Gerald M. Schaflander and Edwin Rothschild, co-founders of Hydrogen Fuels, Inc., 7596 Ventura Canyon Ave., Panorama City, California 91402 (June 1976).

8. "Dr. A.I. Mlavsky Talks About Photovoltaics," interview by John T. Schnebly in *Solar Age* (April 1976).

9. "Business Beat," *Solar Age* (November 1976).

10. Nicholas von Hoffman, "20th Century No Place for Lone Inventor," *The Daily Texan*, Austin, Texas (August 22, 1977).

11. "A New Promise of Cheap Solar Energy," *Business Week* (July 18, 1977).

12. The Adler quote in the *Wall Street Journal* is taken from von Hoffman, *op. cit.*

13. Telephone interview with Dr. Daniel J. Schneider, D.J. Schneider Lift Translator Co., 608 Durango Circle South, Irving, Texas 75062 (February 1979).

14. Documents pertaining to Schneider's experience, including the letter from Stanley I. Weiss, supplied by Schneider; supporting information from Christopher Burke, Assistant Advocate, Energy and Natural Resources, U.S. Small Business Administration, 1441 "L" St. NW, Washington, D.C. 20416 (February 1979).

15. Telephone interview with Gary Nelson, Senior Vice-president, Daystar Corporation, 90 Cambridge St., Burlington, Massachusetts 01803 (August 1977).

16. *Energy Futures, op. cit.*

17. "Business Beat," *Solar Age* (December 1976).

18. Helene Kessler, "The Sun Shines on Wall Street," Pacific News Service, 604 Mission St., Room 1001, San Francisco, California 94105 (July 4, 1977).

19. *Solar Industry Index*, published by Solar Energy Industries Association, 1001 Connecticut Ave. NW, Suite 632, Washington, D.C. 20036 (1977).

20. *Solar Engineering* (May 1977).

21. Information on SEREF from *Solar Engineering* (May 1977); see also Sheldon H. Butt, "Standards for the Solar Industry," *The Solar Market*, proceedings of a symposium sponsored by the Bureau of Competition, Federal Trade Commission, Washington, D.C. (June 1978).

22. "McCormack Reviews Solar Development in National Television Appearance," *Solar Energy Washington Letter* (April 28, 1975).

23. Charles B. Yulish, editor, *Soft vs Hard Energy Paths: 10 Critical Essays on Amory Lovins' "Energy Strategy: The Road Not Taken?"* Charles Yulish Associates, Inc., 229 Seventh Ave., New York, New York 10011 (1977).

24. The letter from McCormack and the Yulish book, with rebuttals from Lovins and other "soft-path" advocates, are reproduced as appendices (1977) to a joint hearing on "Alternative Long-Range Energy Strategies" by the Select Committee on Small Business and the Committee on Interior and Insular Affairs, U.S. Senate, Washington, D.C. (December 9, 1976).

25. Hutchinson's remarks are from Ray Reece, "The Future of Solar Power: Sunrise or Sunset?" *The Texas Observer*, 600 West 7th St., Austin, Texas 78701 (December 19, 1978).

26. "Solar Collector Manufacturing Activity and Applications in the Residential Sector, January Through June, 1977," Energy Information Administration, U.S. Department of Energy, Washington, D.C. (February 1978).

27. Allan Frank, "Flat Plate Collector Manufacturing: Up Again, and Steadying," *Solar Age* (June 1978).

28. *Jobs from the Sun*, California Public Policy Center, Rm 224, 304 S. Broadway, Los Angeles, California 90013 (1978).

CHAPTER XII

1. The quote pertaining to a "husbanding approach toward resources..." taken from the unpublished draft of *Solar Energy in America's Future, op. cit.*

2. Barnet and Muller, *op. cit.*

3. Hayes, *op. cit.*

4. Lovins, *op. cit.*

5. Regarding jobs from solar energy development, see also *Jobs and Energy*, Environmentalists for Full Employment, Room 300, 1785 Massachusetts Ave. NW, Washington, D.C. 20036 (Spring 1977); *Jobs from the Sun*, California Public Policy Center, Los Angeles (1978); *Creating Solar Jobs*, Mid-Peninsula Peace Conversion Project, Mountain View, California (1978); *Long Island Job Study*, James Benson and Steven Buchsbaum for the Council on Economic Priorities, 84 Fifth Ave., New York, New York 10011 (Spring 1979).

6. *Domestic Policy Review of Solar Energy: Response Memorandum to the President*, U.S. Department of Energy (December 3, 1978); see also "Status Report on Solar Energy Domestic Policy Review," Department of Energy (August 25, 1978); *The Great Adventure*, Institute for Local Self-Reliance, Washington, D.C., prepared for U.S. Department of Energy, Washington, D.C. (1978).

7. Lovins' remarks on the DPR from a report in *Energy Daily* (September 21, 1978).

8. *Ways and Means* is the publication of the National Conference on Alternative State and Local Policies, Washington, D.C.

9. For a thorough review of the DPR, see *Inside DOE*, McGraw-Hill, 1221 Avenue of the Americas, New York, New York 10020 (November 27, 1978).

10. The Department of Energy budget for FY 1980 is analyzed in *Energy Daily* (January 23, 1979).

11. HUD and DOE contracts for 1978 are summarized in *Solar Engineering* (May, August, November 1978).

12. *Blueprint for a Solar America*, Solar Lobby, 1028 Connecticut Ave. NW, Washington, D.C. 20036 (December 1978).

13. Lovins quoted in *Energy Daily, op. cit.*

14. "Local/State Energy Efforts," Citizens Energy Project, 1413 "K" St. NW, 8th floor, Washington, D.C. 20005 (1977).

15. *Davis Energy Conservation Report*, Living Systems, Rt. 1, Box 170, Winters, California 95694 (April 1977); this document and others are available from the City of Davis, Community Development Department, 226 "F" St., Davis, California 95616.

16. *Phase II Final Report*, Solar Sustenance Project, Inc., Rt. 1, Box 107AA, Santa Fe, New Mexico 87501 (1977).

17. Susan and Bill Yanda, *The Solar Greenhouse Outreach Program*, Solar Sustenance Project, Inc., *op. cit.* (December 1978).

18. *No Heat, No Rent*, Energy Task Force, 156 Fifth Ave., New York, New York 10010 (1977).

19. *Windmill Power for City People*, Energy Task Force, *op. cit.* (1977).

20. All references to media treatment of the Energy Task Force are taken from a compendium of reprinted articles distributed by ETF on request.